Selected Titles in This Series

(Continued in the back of this publication)

Dynamical Zeta Functions, Nielsen Theory and Reidemeister Torsion

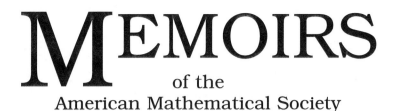

MEMOIRS
of the
American Mathematical Society

Number 699

Dynamical Zeta Functions,
Nielsen Theory and
Reidemeister Torsion

Alexander Fel'shtyn

September 2000 • Volume 147 • Number 699 (third of 4 numbers) • ISSN 0065-9266

American Mathematical Society
Providence, Rhode Island

2000 *Mathematics Subject Classification.*
Primary 58–XX; Secondary 55M20, 57Q10.

Library of Congress Cataloging-in-Publication Data

Fel'shtyn, Alexander, 1952–
 Dynamical zeta functions, Nielsen theory, and Reidemeister torsion / Alexander Fel'shtyn.
  p. cm. — (Memoirs of the American Mathematical Society, ISSN 0065-9266 ; no. 699)
 "Volume 147, number 699 (third of 4 numbers)."
 Includes bibliographical references.
 ISBN 0-8218-2090-7 (alk. paper)
 1. Functions, Zeta. 2. Fixed point theory. 3. Piecewise linear topology. I. Title. II. Series.
QA3 .A57 no. 699
[QA351]
510 s—dc21
[515′.56]   00-034994

Memoirs of the American Mathematical Society

This journal is devoted entirely to research in pure and applied mathematics.

Subscription information. The 2000 subscription begins with volume 143 and consists of six mailings, each containing one or more numbers. Subscription prices for 2000 are $466 list, $419 institutional member. A late charge of 10% of the subscription price will be imposed on orders received from nonmembers after January 1 of the subscription year. Subscribers outside the United States and India must pay a postage surcharge of $30; subscribers in India must pay a postage surcharge of $43. Expedited delivery to destinations in North America $35; elsewhere $130. Each number may be ordered separately; *please specify number* when ordering an individual number. For prices and titles of recently released numbers, see the New Publications sections of the *Notices of the American Mathematical Society*.

Back number information. For back issues see the *AMS Catalog of Publications*.

Subscriptions and orders should be addressed to the American Mathematical Society, P. O. Box 845904, Boston, MA 02284-5904. *All orders must be accompanied by payment.* Other correspondence should be addressed to Box 6248, Providence, RI 02940-6248.

Copying and reprinting. Individual readers of this publication, and nonprofit libraries acting for them, are permitted to make fair use of the material, such as to copy a chapter for use in teaching or research. Permission is granted to quote brief passages from this publication in reviews, provided the customary acknowledgment of the source is given.

Republication, systematic copying, or multiple reproduction of any material in this publication is permitted only under license from the American Mathematical Society. Requests for such permission should be addressed to the Assistant to the Publisher, American Mathematical Society, P. O. Box 6248, Providence, Rhode Island 02940-6248. Requests can also be made by e-mail to `reprint-permission@ams.org`.

Memoirs of the American Mathematical Society is published bimonthly (each volume consisting usually of more than one number) by the American Mathematical Society at 201 Charles Street, Providence, RI 02904-2294. Periodicals postage paid at Providence, RI. Postmaster: Send address changes to Memoirs, American Mathematical Society, P. O. Box 6248, Providence, RI 02940-6248.

Contents

Abstract

In the paper we study new dynamical zeta functions connected with Nielsen fixed point theory. The study of dynamical zeta functions is part of the theory of dynamical systems, but it is also intimately related to algebraic geometry, number theory, topology and statistical mechanics. The paper consists of four parts. Part I presents a brief account of the Nielsen fixed point theory. Part II deals with dynamical zeta functions connected with Nielsen fixed point theory. Part III is concerned with analog of Dold congruences for the Reidemeister and Nielsen numbers. In Part IV we explain how dynamical zeta functions give rise to the Reidemeister torsion , a very important topological invariant which has useful applications in knots theory,quantum field theory and dynamical systems.

Key words and phrases. Dynamical zeta functions, Reidemeister torsion, Nielsen and Reidemeister numbers, fixed point classes and lifting classes, Lefschetz zeta function, Nielsen zeta function, Reidemeister zeta function, functional equation, Dold congruences, topological entropy , Pontryagin duality, space of irreducible unitary representations, Rochlin invariant, topology of an attraction domain.

AMS classification: Primary 58F20; Secondary 55M20, 57Q10.

Introduction

1

0.1 From Riemann zeta function to dynamical zeta functions

In this subsection we shall try to explain where dynamical zeta functions come from . In a sense the study of dynamical zeta functions is part of the theory of dynamical systems, but it is also intimately related to algebraic geometry, number theory, topology and statistical mechanics.

0.1.1 Riemann zeta function

The theory of the Riemann zeta function and its generalisations represent one of the most beatiful developments in mathematics. The Riemann zeta function is that function defined on $\{s \in \mathbb{C} : Re(s) > 1\}$ by the series

$$\zeta(s) \equiv \sum_{n=1}^{\infty} \frac{1}{n^s}.$$

There is a second representation of ζ which was discovered by Euler in 1749 and which is the reason for the significance of the Riemann zeta function in arithmetic. This is Euler product formula:

$$\zeta(s) = \prod_{p \, prime} (1 - p^{-s})^{-1}.$$

---

[1]Received by the editor June 22, 1998

Riemann's significant contribution here was his consideration in 1858 of the zeta function as an analytic function.He first showed that the zeta function has an analytic continuation to the complex plane as a meromorphic function with a single pole at $s = 1$ whose properties can on the one hand be investigated by the techniques of complex analysis, and on the other yield difficult theorems concerning the integers[73]. It is this connection between the continuous and the discrete that is so wonderful. Riemann also showed that the zeta function satisfied a functional equation of the form

$$\zeta(1 - s) = \gamma(s) \cdot \zeta(s)$$

where

$$\gamma(s) = \pi^{1/2 - s} \cdot \Gamma(s/2) / \Gamma((1 - s)/2)$$

and Γ is the Euler gamma-function. In the course of his investigations Riemann was led to suspect that all nontrivial zeros are on the line $Re(s) = 1/2$, this is the Riemann Hypothesis which has been the central goal of research to the present day. Although no proof has yet appeared various weak forms of this conjecture, and in other contexts, analogues of it have been of considerable significance.

0.1.2   Problems concerning zeta functions

Since the 19 th century, many special functions called zeta functions have been defined and investigated. The main problems concerning zeta functions are:

(I) Creation of new zeta functions.

(II) Investigation of the properties of zeta functions. Generally , zeta functions have the following properties in common: 1) They are meromorphic on the whole complex plane ; 2) they have Dirichlet series expansions ; 3) they have Euler product expansions ; 4) they satisfy certain functional equations; 5) their special values play important role. Also, it is an important problem to find the poles, residues, and zeros of zeta functions .

(III) Application to number theory, geometry, dynamical systems.

(IV) Study of the relations between different zeta functions.

Most of the functions called zeta functions or L-functions have the properties of problem (II).

0.1.3 Important types of zeta functions

For a general discussion of a zeta functions see article " Zeta functions " in the Encyclopedic Dictionary of Mathematics [17]. The following is a classification of the important types of zeta functions that are already known:

1) The zeta and L-functions of algebraic numbers fields: the Riemann zeta function, Dirichlet L-functions, Dedekind zeta functions, Hecke L-functions, Artin L-functions.

2) The p-adic L-functions of Leopoldt and Kubota.

3) The zeta functions of quadratic forms: Epstein zeta functions, Siegel zeta functions.

4) The zeta functions associated with Hecke operators.

5) The zeta and L-functions attached to algebraic varieties defined over finite fields: Artin zeta function, Hasse-Weil zeta functions.

6) The zeta functions attached to discontinuous groups : Selberg zeta functions.

7) The dynamical zeta functions: Artin-Mazur zeta function, Lefschetz zeta function, Ruelle zeta function for discrete dynamical systems, Ruelle zeta function for flows.

0.1.4 Hasse-Weil zeta function

Let V be a nonsingular projective algebraic variety of dimension n over a finite field k with q elements. The variety V is thus defined by homogenous polynomial equations with coefficients in the field k for $m+1$ variables $x_0, x_1, ..., x_m$. These variables are in the algebraic closure $\bar{k}$ of the field k, and constitute the homogeneous coordinates of a point of V. The variety V is invariant under the Frobenius map $F : (x_0, x_1, ..., x_m) \rightarrow (x_0^q, x_1^q, ..., x_m^q)$. Arithmetic considerations lead Hasse and Weil to introduce a zeta function which counts the points of V with coordinates in the different finite extensions of the field k, or equivalently points of V which are fixed under F^n for some $n \geq 1$:

$$\zeta(z, V) := \exp\left(\sum_{n=1}^{\infty} \frac{\#\text{Fix}\,(F^n)}{n} z^n\right)$$

Note that $\zeta(z, V)$ can be written as a Euler product

$$\zeta(z, V) = \prod_{\gamma} \frac{1}{1 - z^{\#\gamma}},$$

over all primitive periodic orbits γ of F on V. For comparison with Riemann's zeta function one has to put $z = q^{-s}$. Certain conjectures proposed by Weil [98] on the properties of $\zeta(z, V)$ led to lot of work by Weil, Dwork, Grothendieck, and complete proof was finally obtained by Deligne [15]. The story of the Weil conjectures is one of the most striking instances exhibiting the fundamental unity of mathematics. It is found that $\zeta(z, V)$ is a rational function of z with a functional equation :

$$\zeta(z, V) = \prod_{i=0}^{2m} P_l(z)^{(-1)^{l+1}}$$

where the zeros of the polynomial P_l have absolute value $q^{-l/2}$ and the $P_l(z)$ have a cohomological interpretation: the polynomial P_l is roughly the characteristic polynomial associated with the induced action of the Frobenius morphism on the etale cohomology: $P_l(z) = \det(1 - z \cdot F^* | H^l(V))$.

0.1.5   Dynamical zeta functions

Inspired by the Hasse-Weil zeta function of an algebraic variety over a finite field, Artin and Mazur [5] defined the Artin - Mazur zeta function for an arbitrary map $f : X \to X$ of a topological space X:

$$F_f(z) := \exp\left(\sum_{n=1}^{\infty} \frac{F(f^n)}{n} z^n\right)$$

where $F(f^n)$ is the number of isolated fixed points of f^n. Artin and Mazur showed that for a dense set of the space of smooth maps of a compact smooth manifold into itself the Artin-Mazur zeta function $F_f(z)$ has a positive radius of convergence.Later Manning [64] proved the rationality of the Artin - Mazur zeta function for diffeomorphisms of a smooth compact manifold satisfying Smale axiom A, after partial results were obtained by Williams and Guckenheimer. On the other hand there exist maps for which Artin-Mazur zeta function is transcendental [11].

The Artin-Mazur zeta function was adopted later by Milnor and Thurston [67] to count periodic points for a piecewise monotone map of the interval.

The Artin-Mazur zeta function was historically the first dynamical zeta function for *discrete* dynamical system. The next dynamical zeta function was defined by Smale [87] .This is the Lefschetz zeta function of discrete dynamical system:

$$L_f(z) := \exp\left(\sum_{n=1}^{\infty} \frac{L(f^n)}{n} z^n\right),$$

where

$$L(f^n) := \sum_{k=0}^{\dim X} (-1)^k \mathrm{Tr}\left[f_{*k}^n : H_k(X; \mathbb{Q}) \to H_k(X; \mathbb{Q})\right]$$

is the Lefschetz number of f^n. Smale considered $L_f(z)$ in the case when f is diffeomorphism of a compact manifold, but it is well defined for any continuous map f of compact polyhedron X. The Lefschetz zeta function is a rational function of z and is given by the formula:

$$L_f(z) = \prod_{k=0}^{\dim X} \det\left(I - f_{*k} \cdot z\right)^{(-1)^{k+1}}.$$

Afterwards, J.Franks [38] defined reduced mod 2 Artin-Mazur and Lefschetz zeta functions, and D. Fried [40] defined twisted Artin-Mazur and Lefschetz zeta functions, which have coefficients in the group rings $\mathbb{Z}H$ or $\mathbb{Z}_2 H$ of an abelian group H. The above zeta functions are directly analogous to the Hasse-Weil zeta function.

Ruelle has found another generalization of the Artin-Mazur zeta function. He was motivated by ideas from equilibrium statistical mechanics and has replaced in the Artin-Mazur zeta function [82] simple counting of the periodic points by counting with weights. He defined the Ruelle zeta function as

$$F_f{}^g(z) := \exp\left(\sum_{n=1}^{\infty} \frac{z^n}{n} \sum_{x \in \mathrm{Fix}\ (f^n)} \prod_{k=0}^{n-1} g(f^k(x))\right),$$

where $g : X \to \mathbb{C}$ is a weight function(if $g = 1$ we recover $F_f(z)$).

Dynamical zeta functions have relations with statistical mechanics(entropy, pressure, Gibbs states, equilibrium states). Manning used Markov partitions and corresponding symbolic dynamics in his proof of the rationality of the Artin-Mazur zeta function. This symbolic dynamics is reminiscent of the statistical mechanics of one-dimensional lattice spin system.

0.2 Dynamical zeta functions and Nielsen fixed point theory

In contrast with the Artin- Mazur zeta function which counts the periodic points of the map geometrically, the Lefschetz zeta function does this algebraically.There is another way of counting the fixed points of f^n - according to Nielsen [51].

Let $p : \tilde{X} \to X$ be the universal covering of X and $\tilde{f} : \tilde{X} \to \tilde{X}$ a lifting of f, ie. $p \circ \tilde{f} = f \circ p$. Two liftings $\tilde{f}$ and $\tilde{f}'$ are called *conjugate* if there is a $\gamma \in \Gamma \cong \pi_1(X)$ such that $\tilde{f}' = \gamma \circ \tilde{f} \circ \gamma^{-1}$. The subset $p(Fix(\tilde{f})) \subset Fix(f)$ is called *the fixed point class of f determined by the lifting class $[\tilde{f}]$*. A fixed point class is called *essential* if its index is nonzero. The number of lifting classes of f (and hence the number of fixed point classes, empty or not) is called the *Reidemeister Number* of f, denoted $R(f)$. This is a positive integer or infinity. The number of essential fixed point classes is called the *Nielsen number* of f, denoted by $N(f)$. The Nielsen number is always finite. $R(f)$ and $N(f)$ are homotopy type invariants. In the category of compact, connected polyhedra the Nielsen number of a map is equal to the least number of fixed points of maps with the same homotopy type as f.In Nielsen fixed point theory the main objects for investigation are the Nielsen and Reidemeister numbers and their modifications [51].

Let G be a group and $\phi : G \to G$ an endomorphism. Two elements $\alpha, \alpha' \in G$ are said to be $\phi - conjugate$ iff there exists $\gamma \in G$ such that $\alpha' = \gamma.\alpha.\phi(\gamma)^{-1}$. The number of ϕ-conjugacy classes is called the *Reidemeister number* of ϕ, denoted by $R(\phi)$.

In papers [23, 24, 25, 28] we have introduced new dynamical zeta functions connected with Nielsen fixed point theory.We defined the Nielsen zeta function of f and Reidemeister zeta functions of f and ϕ as power series:

$$R_\phi(z) := \exp\left(\sum_{n=1}^{\infty} \frac{R(\phi^n)}{n} z^n\right),$$

$$R_f(z) := \exp\left(\sum_{n=1}^{\infty} \frac{R(f^n)}{n} z^n\right),$$

$$N_f(z) := \exp\left(\sum_{n=1}^{\infty} \frac{N(f^n)}{n} z^n\right).$$

We assume that $R(f^n) < \infty$ and $R(\phi^n) < \infty$ for all $n > 0$.

We have investigated the following problem: for which spaces and maps and for which groups and endomorphisms are the Nielsen and Reidemeister zeta functions a rational functions?When do they have a functional equation? Are these functions algebraic functions?

In [23] we proved that the Nielsen zeta function has a positive radius of convergence which admits a sharp estimate in terms of the topological entropy of the map. Later, in [35] we propose another prove of positivity of radius and proved an exact algebraic lower estimation for it. With the help of Nielsen - Thurston theory [46] of surface homeomorphisms , in [74] we proved that for an orientation-preserving homeomorphism of a compact surface the Nielsen zeta function is either a rational function or the radical of rational function. For a periodic map of any compact polyhedron in [74] we proved a product formula for Nielsen zeta function which implies that Nielsen zeta function is a radical of a rational function.

The investigation and computation of the Reidemeister zeta function $R_\phi(z)$of a group endomorphism ϕ is an algebraic ground of the computation and investigation of zeta functions $R_f(z)$ and $N_f(z)$. In [25] we investigated the behavior of $R_\phi(z)$under the extension of a group and proved rationality and a functional equation of $R_\phi(z)$ and a trace formula for the $R(\phi^n)$ for the endomorphism of a finitely generated free Abelian group and group $\mathbb{Z}/p\mathbb{Z}$. An endomorphism $\phi : G \to G$ is said to be eventually commutative if there exists a natural number n such that the subgroup $\phi^n(G)$ is commutative. A map $f : X \to X$ is said to be eventually commutative if the induced endomorphism on fundamental group is eventually commutative. In [31, 32, 33] we proved that $R_\phi(z)$ is a rational function with functional equation in the case of any endomorphism ϕ of any finite group G and in the case that ϕ is eventually commutative and G is finitely generated. As a consequence we obtained rationality and a functional equation for $R_f(z)$ where either the fundamental group of X is finite, or the map f is eventually commutative.We obtained sufficient conditions under which the Nielsen zeta function coincides with the Reidemeister zeta function and is a rational function with functional equation. As an application we calculate the Reidemeister and Nielsen zeta functions of all self-maps of lens spaces, nilmanifolds and tori.

In [24, 25, 27] we found a connection between the rationality of the Nielsen and Reidemeister zeta functions for the maps of fiber,base and total space of a fiber map of a Serre bundle using results of Brown, Fadell, and You (see [51]) about Nielsen and Reidemeister numbers of a fiber map.

In [34] we proved the rationality of the Reidemeister zeta function and the trace formulas for the Reidemeister numbers of group endomorphisms in the following cases: the group is finitely generated and an endomorphism is eventually commutative; the group is finite ; the group is a direct sum of a finite group and a finitely generated free Abelian group; the group is finitely generated, nilpotent and torsion free .

0.3   Congruences for Reidemeister numbers

In his article [16], Dold found a remarkable arithmetical property of the Lefschetz numbers for the iterations of a map f. He proved the following formula

$$\sum_{d|n} \mu(d) \cdot L(f^{n/d}) \equiv 0 \bmod n$$

where n is any natural number and μ is the Möbius function. This result had previously been obtained for prime n by Zabreiko, Krasnosel'skii [102] and Steinlein [89].

The congruences for Lefschetz numbers are directly connected with the rationality of the Lefschetz zeta function [16].

In [31], [35] we proved, under additional conditions, similar congruences :

$$\sum_{d|n} \mu(d) \cdot R(\phi^{n/d}) \equiv 0 \bmod n,$$

$$\sum_{d|n} \mu(d) \cdot R(f^{n/d}) \equiv 0 \bmod n$$

for the Reidemeister numbers of the iterations of a group endomorphism ϕ and a map f.

This result implies , in special cases , the corresponding congruences for the Nielsen numbers. For n-toral maps in the case when Jiang subgroup coincide with fundamental group this congruences for Nielsen numbers were proved by Heath, Piccinini, and You [49] .

In [36], we generalize the arithmetic congruence relations among the Reidemeister numbers of iterates of maps to similar congruences for Reidemeister numbers of equivariant group endomorphisms and maps.

In the article [32] we conjected a general connection between the Reidemeister number of a group endomorphism and the number of fixed points of the induced map on the space of irreducible unitary representations. This can be reformulated in terms of self-maps and pullbacks of vector bundles. The results [32, 33] are essentially proofs of these conjectures under the condition that ϕ is eventually commutative or G is finite.

0.4 Reidemeister torsion.

Dynamical zeta functions in the Nielsen theory are closely connected with the Reidemeister torsion.

Reidemeister torsion is a very important topological invariant which has useful applications in knots theory,quantum field theory and dynamical systems.In 1935 Reidemeister [78] classified up to PL equivalence the lens spaces S^3/Γ where Γ is a finite cyclic group of fixed point free orthogonal transformations. He used a certain new invariant which was quickly extended by Franz , who used it to classify the generalized lens spaces S^{2n+1}/Γ. This invariant is a ratio of determinants concocted from a Γ-equivariant chain complex of S^{2n+1} and a nontrivial character $\rho : \Gamma \to U(1)$ of Γ. Such a ρ determines a flat bundle E over S^{2n+1}/Γ such that E has holonomy ρ. The new invariant is now called the *Reidemeister torsion*, or *R-torsion* of E.

The results of Reidemeister and Franz were extended by de Rham to spaces of constant curvature $+1$.

Whitehead refined and generalized Reidemeister torsion in defining the torsion of a homotopy equivalence in 1950, and his work was to play a crucial role in the development of geometric topology and algebraic K-theory in the 60's. The Reidemeister torsion is closely related to the K_1 groups of algebraic K-theory.

Later Milnor identified the Reidemeister torsion with the Alexander polynomial, which plays a fundamental role in the theory of knots and links.

In 1971, Ray and Singer [77] introduced an analytic torsion associated with the de Rham complex of forms with coefficients in a flat bundle over a compact Riemannian manifold, and conjectured it was the same as the Reidemeister torsion associated with the action of the fundamental group on the covering space, and the representation associated with the flat bundle. The Ray- Singer conjecture was established independently by Cheeger [13] and Müller [69] a few years later.

In 1978 A.Schwartz [84] showed how to construct a quantum field theory on a manifold M whose partition function is a power of the analytical torsion of M. Witten [101] has used analytical torsion to study non-Abelian Chern-Simons gauge field theory. Its partition function is the Witten-Reshetikhin-Turaev invariant[95] for the three manifold M and the analytic torsion appears naturally in the asymptotic formula for the partition function obtained by the method of stationary phase approximation [101].

Recently, the Reidemeister torsion has found interesting applications in dynamical systems theory. A connection between the Lefschetz type dynamical zeta functions and the Reidemeister torsion was established by D. Fried [40]. The work of Milnor [65] was the first indication that such a connection exists. Fried also has shown that, for some flows, the value at 0 of the Ruelle zeta function coincides with the Reidemeister torsion.

In [26], [32, 33, 34] we established a connection between the Reidemeister torsion and Reidemeister zeta function. We obtained an expression for the Reidemeister torsion of the mapping torus of the dual map of a group endomorphism, in terms of the Reidemeister zeta function of the endomorphism. The result is obtained by expressing the Reidemeister zeta function in terms of the Lefschetz zeta function of the dual map, and then applying the theorem of D. Fried. What this means is that the Reidemeister torsion counts the fixed point classes of all iterates of map f i.e. periodic point classes of f.

In [29] we established a connection between the Reidemeister torsion of a mapping torus, the eta-invariant, the Rochlin invariant and the multipliers of the theta function. The formula is obtained via the Lefschetz zeta function and the results on the holonomy of determinant line bundles due to Witten [100], Bismut-Freed [9], and Lee, Miller and Weintraub [61].

Note, that the work of Turaev [94] was the first indication that the Rochlin invariant is connected with the Reidemeister torsion for three-dimensional rational homology spheres.

In [43], we proved an analogue of the Morse inequalities for the attraction domain of an attractor, and the level surface of the Lyapunov function. These inequalities describe the connection between the topology of the attraction domain and dynamics of a Morse-Smale flow on the attractor.

In [26, 30] we described with the help of the Reidemeister torsion the connection between the topology of the attraction domain of an attractor and the dynamic of flow with circular chain-recurrent set on the attractor. We showed that for flow with circular chain-recurrent set, the Reidemeister torsion of the attraction domain of an attractor and Reidemeister torsion of the level surface of the Lyapunov function is a special value of the twisted Lefschetz zeta function building via closed orbits in the attractor.

In [30] we found that for the integrable Hamiltonian system on the four-dimensional symplectic manifold, the Reidemeister torsion of the isoenergetic surface counts the critical circles (which are the closed trajectories of the system) of the second independent Bott integral on this surface.

0.5   Table of contents

The monograph consists of four parts. Part I(Chapter 1) presents a brief account of the Nielsen fixed point theory. Part II(Chapters 2 - 4) deals with dynamical zeta functions connected with Nielsen fixed point theory. Part III (Chapter 5) is concerned with congruences for the Reidemeister and Nielsen numbers. Part IV (Chapter 6) deals with the Reidemeister torsion .

The content of the chapters should be clear from the headings. The following remarks give more directions to the reader.

In Chapter 1 we define the lifting and fixed point classes, fixed point index, Reidemeister and Nielsen numbers. The relevant definitions and results will be used throughout the book.

In Chapter 2 - 4 we introduce the Reidemeister zeta functions of a group endomorphism and of a map and the Nielsen zeta function of a map which are the main objects of the monograph.

In Chapter 2 we prove that the Reidemeister zeta function of a group endomorphism is a rational function with functional equation in the following cases: the group is finitely generated and an endomorphism is eventually commutative; the group is finite ; the group is a direct sum of a finite group and a finitely generated free abelian group; the group is finitely generated, nilpotent and torsion free. As a consequence we obtained rationality and a functional equation for the Reidemeister zeta function of a continuous map where the fundamental group of X is as above.

In Chapter 3 we show that the Nielsen zeta function has a positive radius of convergence which admits a sharp estimate in terms of the topological entropy of the map. We also give an exact algebraic lower estimation for the radius.With the help of Nielsen - Thurston theory of surface homeomorphisms we prove that for an orientation-preserving homeomorphism of a compact surface the Nielsen zeta function is either a rational function or the radical of rational function. For a periodic map of a compact polyhedron we prove a product formula for Nielsen zeta function which implies that Nielsen zeta function is a radical of a rational function. In section 3.4 and 3.5 we give sufficient conditions under which the Nielsen zeta function coincides with the Reidemeister zeta function and is a rational function with functional equation. In section 3.6 we describe connection between the rationality of the Nielsen zeta functions for the maps of fiber, base and total space of a fiber map of a Serre bundle. We would like to mention that in all known cases the Nielsen zeta function is a nice function. By this we mean that it is a product of an exponential of a polynomial with a function some power of which is rational. May be this is a general pattern.

In Chapter 4 we generalize the results of Chapter 2-3 to the Nielsen and Reidemeister zeta functions modulo normal subgroup of the fundamental group.

In Chapter 5 we prove analog of Dold congruences for Reidemeister and Nielsen numbers.

In Chapter 6 we explain how dynamical zeta functions give rise to the Reidemeister torsion, a very important topological invariant . In section 6.2 we establish a connection between the Reidemeister torsion and Reidemeister zeta function. We obtain an expression for the Reidemeister torsion of the mapping torus of the dual map of a group endomorphism, in terms of the Reidemeister zeta function of the endomorphism.This means that the Reidemeister torsion counts the fixed point classes of all iterates of map f i.e. periodic point classes of f.

In section 6.3 we establish a connection between the Reidemeister torsion of a mapping torus, the eta-invariant, the Rochlin invariant and the multipliers of the theta function.

In section 6.4 we describe with the help of the Reidemeister torsion and of an analog of Morse inequalities the connection between the topology of the attraction domain of an attractor and the dynamics of the system on the attractor.

In section 6.5 we show that for the integrable Hamiltonian system on the four-dimensional symplectic manifold, the Reidemeister torsion of the isoenergetic surface counts the critical circles(which are the closed trajectories of the system) of the second independent Bott integral on this surface.

A part of this monograph grew out of the joint papers of the author with V.B. Pilyugina and R. Hill written in 1985-1995. Their collaboration is gratefully appreciated.

This book had its beginnings in talks which the author gave at Rochlin seminar in 1983 - 1990. We are happy to acknowledge the influence of V.A. Rochlin on our approach to the subject of the book.

The author would like to thank D. Anosov, A. Dold, J. Eichhorn, D. Fried, M.Gromov, B.Jiang, S. Patterson, A. Ranicki, V.Turaev, O. Viro, P. Wong for useful discussions and comments.

Parts of this book were written while the author was visiting the University of Göttingen, Institute des Hautes Etudes Scientifiques (Bures-sur - Yvette), Max-Planck-Institut für Mathematik (Bonn).The author is indebted to these institutions for their invitations and hospitality.

Chapter 1

Nielsen Fixed Point Theory

In this section we give a brief review of the Nielsen theory.

1.1 History

Fixed point theory started in the early days of topology, because of its close relationship with other branches of mathematics.Existence theorems are often proved by converting the problem into an appropriate fixed point problem.Examples are the existence of solutions for elliptic partial differential equations, and the existence of closed orbits in dynamical systems. In many problems, however, one is not satisfied with the mere existence of a solution. One wants to know the number, or at least a lower bound for the number of solutions. But the actual number of fixed points of a map can hardly be the subject of an interesting theory, since it can be altered by an arbitrarily small perturbation of the map. So, in topology, one proposes to determine the minimal number of fixed points in a homotopy class.This is what Nielsen fixed point theory about. Perhaps the best known fixed point theorem in topology is the Lefschetz fixed point theorem

Theorem 1 *[62] Let X be a compact polyhedron, and $f : X \to X$ be a map. If the Lefschetz number $L(f) \neq 0$, then every map homotopic to f has a fixed point.*

The Lefschetz number is the total algebraic count of fixed points.It is a homotopy invariant and is easily computable.

14

So, the Lefschetz theorem, along with its special case, the Brouwer fixed point theorem, and its generalization, the widely used Leray-Schauder theorem in functional analysis, can tell existence only. In contrast, the chronologically first result of Nielsen theory has set a beautiful example of a different type of theorem

Theorem 2 *[70] Let $f : T^2 \to T^2$ be a map of the torus. Suppose that the endomorphism induced by f on the fundamental group $\pi_1(T^2) \cong Z \oplus Z$ is represented by the 2×2 integral matrix A. Then the least number of fixed points in the homotopy class of f equals the absolute value of the determinant of $E - A$, where E is the identity matrix, i.e.*

$$Min\{\#\text{Fix}\,(g) \mid g \simeq f\} = |\det(E - A)|.$$

It can be shown that $\det(E - A)$ is exactly $L(f)$ on tori and $|\det(E - A)|$ is the Nielsen number $N(f)$ on tori. This latter theorem says much more than the Lefschetz theorem specialised to the torus, since it gives a lower bound for the number of fixed points, or it confirms the existence of a homotopic map which is fixed point free. The proof was via the universal covering space R^2 of the torus. From this instance evolved the central notions of Nielsen theory - the fixed point classes and the Nielsen number.

Roughly speaking, Nielsen theory has two aspects. The geometric aspect concerns the comparison of the Nielsen number with the least number of fixed points in a homotopy class of maps. The algebraic aspect deals with the problem of computation for the Nielsen number. Nielsen theory is based on the theory of covering spaces. An alternative way is to consider nonempty fixed point classes only, and use paths instead of covering spaces to define them. This is certainly more convenient for some geometric questions. But the covering space approach is theoretically more satisfactory, especially for computational problems, since the nonemptiness of certain fixed point classes is often the conclusion of the analysis, not the assumption.

Now let us introduce the basic idea of Nielsen theory by an elementary example(see [51])

Example 1 *Let $f : S^1 \to S^1$ be a map of the circle. Suppose the degree of f is d. Then the least number of fixed points in the homotopy class of f is $|1 - d|$.*

PROOF Let S^1 be the unit circle on the complex plane, i.e. $S^1 = \{z \in C \mid |z| = 1\}$. Let $p : R \to S^1$ be the exponential map $p(\theta) = z = e^{i\theta}$. Then θ is the argument of z, which is multi-valued function of z. For every $f : S^1 \to S^1$, one can always find "argument expressions" (or liftings) $\tilde{f} : R \to R$ such that $f(e^{i\theta}) = e^{i\tilde{f}(\theta)}$, in fact a whole series of them, differing from each other by integral multiples of 2π. For definiteness let us write $\tilde{f}_0$ for the argument expression with $\tilde{f}_0(0)$ lying in $[0, 2\pi)$, and write $\tilde{f}_k = \tilde{f}_0 + 2k\pi$. Since the degree of f is d, the functions $\tilde{f}_k$ are such that $\tilde{f}_k(\theta + 2\pi) = \tilde{f}_k(\theta) + 2d\pi$. For example, if $f(z) = -z^d$, then $\tilde{f}_k(\theta) = d\theta + (2k + 1)\pi$. It is evident that if $z = e^{i\theta}$ is a fixed point of f, i.e. $z = f(z)$, then θ is a fixed point of some argument expression of f, i.e. $\theta = \tilde{f}_k(\theta)$ for some k. On the other hand, if θ is a fixed point of $\tilde{f}_k$, q is an integer, then $\theta + 2q\pi$ is a fixed point of $\tilde{f}_l$ iff $l - k = q(1 - d)$. This follows from the calculation

$$\tilde{f}_l(\theta + 2q\pi) = \tilde{f}_k(\theta + 2q\pi) + 2(l - k)\pi = \tilde{f}_k(\theta) + 2qd\pi + 2(l - k)\pi =$$

$$= (\theta + 2q\pi) + 2\pi\{(l - k) - q(1 - d)\}.$$

Thus , if $l \not\equiv k \mod (1 - d)$, then a fixed point of $\tilde{f}_k$ and a fixed point of $\tilde{f}_l$ can never correspond to the same fixed point of f , i.e. $p(\text{Fix } (\tilde{f}_k)) \cap p(\text{Fix } (\tilde{f}_l)) = \emptyset$. So, the argument expressions fall into equivalence classes (called lifting classes) by the relation $\tilde{f}_l \sim \tilde{f}_k$ iff $k \equiv l \mod (1 - d)$, and the fixed points of f split into $|1 - d|$ classes (called fixed point classes) of the form $p(\text{Fix } (\tilde{f}_k))$. That is , two fixed points are in the same class iff they come from fixed points of the same argument expression. Note that each fixed point class is by definition associated with a lifting class, so that the number of fixed point classes is $|1 - d|$ if $d \neq 1$ and is ∞ if $d = 1$. Also note that a fixed point class need not be nonempty. Now, to prove that a map f of degree d has at least $|1 - d|$ fixed points, we only have to show that every fixed point class is nonempty, or equivalently, that every argument expression has a fixed point, if $d \neq 1$. In fact, for each k, by means of the equality $\tilde{f}_k(\theta + 2\pi) - \tilde{f}_k(\theta) = 2d\pi$, it is easily seen that the function $\theta - \tilde{f}_k(\theta)$ takes different signs when θ approaches $\pm\infty$, hence $\tilde{f}_k(\theta)$ has at least one fixed point. That $|1 - d|$ is indeed the least number of fixed points in the homotopy class is seen by checking the special map $f(z) = -z^d$.

The following sections can be considered as generalization of this simplest example.

1.2   Lifting classes and fixed point classes

Let $f : X \to X$ be a continuous map of a compact connected polyhedron. Let $p : \tilde{X} \to X$ be the universal covering of X. A lifting of f is a map $\tilde{f} : \tilde{X} \to \tilde{X}$ such that $p \circ \tilde{f} = f \circ p$. A covering translation is a map $\gamma : \tilde{X} \to \tilde{X}$ such that $p \circ \gamma = p$, i.e. a lifting of the identity map. Now we describe standard facts from covering space theory

Proposition 1 *(1) For any $x_0 \in X$ and any $\tilde{x}_0, \tilde{x}_0' \in p^{-1}(x_0)$, there is a unique covering translation $\gamma : \tilde{X} \to \tilde{X}$ such that $\gamma(\tilde{x}_0) = \tilde{x}_0'$. The covering translations of $\tilde{X}$ form a group Γ which is isomorphic to $\pi_1(X)$.*
(2) Let $f : X \to X$ be a continuous map. For given $x_0 \in X$ and $x_1 = f(x_0)$, pick $\tilde{x}_0 \in p^{-1}(x_0)$ and $\tilde{x}_1 \in p^{-1}(x_1)$ arbitrarily. Then, there is a unique lifting of f such that $\tilde{f}(\tilde{x}_0) = \tilde{x}_1$.
(3) Suppose $\tilde{f}$ is a lifting of f, and $\alpha, \beta \in \Gamma$. Then $\beta \circ \tilde{f} \circ \alpha$ is a lifting of f.
(4) For any two liftings $\tilde{f}$ and $\tilde{f}'$ of f, there is a unique $\gamma \in \Gamma$, such that $\tilde{f}' = \gamma \circ \tilde{f}$.

Lemma 1 *Suppose $\tilde{x} \in p^{-1}(x)$ is a fixed point of lifting $\tilde{f}$ of f, and $\gamma \in \Gamma$ is a covering translation on $\tilde{X}$. Then, a lifting $\tilde{f}'$ of f has $\gamma(\tilde{x}) \in p^{-1}(x)$ as a fixed point iff $\tilde{f}' = \gamma \circ \tilde{f} \circ \gamma^{-1}$.*

PROOF "If" is obvious : $\tilde{f}'(\gamma(\tilde{x})) = \gamma \circ \tilde{f} \circ \gamma^{-1}(\gamma(\tilde{x})) = \gamma \circ \tilde{f}(\tilde{x}) = \gamma(\tilde{x})$. " Only if" : Both $\tilde{f}'$ and $\gamma \circ \tilde{f} \circ \gamma^{-1}$ have $\gamma(\tilde{x})$ as a fixed point , so they agree at the point $\gamma(\tilde{x})$. By Proposition 1 they are the same lifting.

Definition 1 *Two liftings $\tilde{f}'$ and $\tilde{f}$ of f are said to be conjugate if there exists $\gamma \in \Gamma$, such that $\tilde{f}' = \gamma \circ \tilde{f} \circ \gamma^{-1}$. Lifting classes are equivalence classes by conjugacy. Notation:*

$$[\tilde{f}] = \{\gamma \circ \tilde{f} \circ \gamma^{-1} \mid \gamma \in \Gamma\}$$

Lemma 2 *(1) Fix $(f) = \cup_{\tilde{f}} p(\text{Fix } (\tilde{f}))$.*
(2) $p(\text{Fix } (\tilde{f})) = p(\text{Fix } (\tilde{f}'))$ if $[\tilde{f}] = [\tilde{f}']$.
(3) $p(\text{Fix } (\tilde{f})) \cap p(\text{Fix } (\tilde{f}')) = \emptyset$ if $[\tilde{f}] \neq [\tilde{f}']$.

PROOF   (1) If $x_0 \in$ Fix (f), pick $\tilde{x}_0 \in p^{-1}(x_0)$. By proposition 1 there exists $\tilde{f}$ such that $\tilde{f}(\tilde{x}_0) = \tilde{x}_0$. Hence $x_0 \in p(\text{Fix } (\tilde{f}))$.

(2) If $\tilde{f}' = \gamma \circ \tilde{f} \circ \gamma^{-1}$, then by lemma Fix $(\tilde{f}') = \gamma$Fix $(\tilde{f})$, so $p(\text{Fix }(\tilde{f})) = p(\text{Fix }(\tilde{f}'))$.

(3) If $x_0 \in p(\text{Fix }(\tilde{f})) \cap p(\text{Fix }(\tilde{f}'))$, there are $\tilde{x}_0, \tilde{x}_0' \in p^{-1}(x_0)$ such that $\tilde{x}_0 \in \text{Fix }(\tilde{f})$ and $\tilde{x}_0' \in \text{Fix }(\tilde{f}')$. Suppose $\tilde{x}_0' = \gamma \tilde{x}_0$. By lemma , $\tilde{f}' = \gamma \circ \tilde{f} \circ \gamma^{-1}$, hence $[\tilde{f}] = [\tilde{f}']$.

Definition 2 *The subset $p(\text{Fix }(\tilde{f}))$ of Fix (f) is called the fixed point class of f determined by the lifting class $[\tilde{f}]$.*

We see that the fixed point set Fix (f) splits into disjoint union of fixed point classes.

Example 2 *Let us consider the identity map $id_X : X \to X$. Then a lifting class is a usual conjugasy class in Γ; $p(\text{Fix }(id_{\tilde{X}})) = X$ and $p(\text{Fix }(\gamma)) = \emptyset$ otherwise.*

Remark 1 *A fixed point class is always considered to carry a label - the lifting class determining it . Thus two empty fixed point classes are considered different if they are determined by different lifting classes.*

Our definition of a fixed point class is via the universal covering space. It essentially says: Two fixed point of f are in the same class iff there is a lifting $\tilde{f}$ of f having fixed points above both of them. There is another way of saying this, which does not use covering space explicitly, hence is very useful in identifying fixed point classes.

Lemma 3 ([51]) *Two fixed points x_0 and x_1 of f belong to the same fixed point class iff there is a path c from x_0 to x_1 such that $c \cong f \circ c$ (homotopy relative endpoints).*

Lemma 3 can be considered as an equivalent definition of a non-empty fixed point class. Every map f has only finitely many non-empty fixed point classes, each a compact subset of X.

1.2.1   The influence of a homotopy

Given a homotopy $H = \{h_t\} : f_0 \cong f_1$, we want to see its influence on fixed point classes of f_0 and f_1. A homotopy $\tilde{H} = \{\tilde{h}_t\} : \tilde{X} \to \tilde{X}$ is called a lifting of the homotopy $H = \{h_t\}$, if $\tilde{h}_t$ is a lifting of h_t for every $t \in I$. Given a homotopy H and a lifting $\tilde{f}_0$ of f_0, there is a unique lifting $\tilde{H}$ of H such that $\tilde{h}_0 = \tilde{f}_0$, hence by unique lifting property of covering spaces they determine a lifting $\tilde{f}_1$ of f_1. Thus H gives rise to a one-one correspondence from liftings of f_0 to liftings of f_1. This correspondence preserves the conjugacy relation.Thus there is a one-to-one correspondence between lifting classes and fixed point classes of f_0 and those of f_1.

1.3   Reidemeister numbers

1.3.1   Reidemeister numbers of a continuous map

Definition 3 *The number of lifting classes of f (and hence the number of fixed point classes, empty or not) is called the Reidemeister number of f, denoted by $R(f)$. It is a positive integer or infinity.*

The Reidemeister number $R(f)$ is a homotopy invariant.

Example 3 *If X is simply-connected then $R(f) = 1$.*

Let $f : X \to X$ be given, and let a specific lifting $\tilde{f} : \tilde{X} \to \tilde{X}$ be chosen as reference. Let Γ be the group of covering translations of $\tilde{X}$ over X. Then every lifting of f can be written uniquely as $\alpha \circ \tilde{f}$, with $\alpha \in \Gamma$. So elements of Γ serve as coordinates of liftings with respect to the reference $\tilde{f}$. Now for every $\alpha \in \Gamma$ the composition $\tilde{f} \circ \alpha$ is a lifting of f so there is a unique $\alpha' \in \Gamma$ such that $\alpha' \circ \tilde{f} = \tilde{f} \circ \alpha$. This correspondence $\alpha \to \alpha'$ is determined by the reference $\tilde{f}$, and is obviously a homomorphism.

Definition 4 *The endomorphism $\tilde{f}_* : \Gamma \to \Gamma$ determined by the lifting $\tilde{f}$ of f is defined by*

$$\tilde{f}_*(\alpha) \circ \tilde{f} = \tilde{f} \circ \alpha.$$

It is well known that $\Gamma \cong \pi_1(X)$. We shall identify $\pi = \pi_1(X, x_0)$ and Γ in the following way. Pick base points $x_0 \in X$ and $\tilde{x}_0 \in p^{-1}(x_0) \subset \tilde{X}$ once and for all. Now points of $\tilde{X}$ are in 1-1 correspondence with homotopy classes of paths in X which start at x_0: for $\tilde{x} \in \tilde{X}$ take any path in $\tilde{X}$ from $\tilde{x}_0$ to $\tilde{x}$ and project it onto X; conversely for a path c starting at x_0, lift it to a path in $\tilde{X}$ which starts at $\tilde{x}_0$, and then take its endpoint. In this way, we identify a point of $\tilde{X}$ with a path class $< c >$ in X starting from x_0. Under this identification, $\tilde{x}_0 =< e >$ is the unit element in $\pi_1(X, x_0)$. The action of the loop class $\alpha =< a >\in \pi_1(X, x_0)$ on $\tilde{X}$ is then given by

$$\alpha =< a >:< c >\longrightarrow \alpha.c =< a.c > .$$

Now we have the following relationship between $\tilde{f}_* : \pi \to \pi$ and

$$f_* : \pi_1(X, x_0) \longrightarrow \pi_1(X, f(x_0)).$$

Lemma 4 ([51]) *Suppose $\tilde{f}(\tilde{x}_0) =< w >$. Then the following diagram commutes:*

$$\begin{array}{ccc}
\pi_1(X, x_0) & \xrightarrow{\;f_*\;} & \pi_1(X, f(x_0)) \\
_{\tilde{f}_*}\searrow & & \downarrow w_* \\
& & \pi_1(X, x_0)
\end{array}$$

We have seen that $\alpha \in \pi$ can be considered as the coordinate of the lifting $\alpha \circ \tilde{f}$. Can we tell the conjugacy of two liftings from their coordinates?

Lemma 5 $[\alpha \circ \tilde{f}] = [\alpha' \circ \tilde{f}]$ *iff there is $\gamma \in \pi$ such that $\alpha' = \gamma\alpha\tilde{f}_*(\gamma^{-1})$.*

PROOF $[\alpha \circ \tilde{f}] = [\alpha' \circ \tilde{f}]$ iff there is $\gamma \in \pi$ such that $\alpha' \circ \tilde{f} = \gamma \circ (\alpha \circ \tilde{f}) \circ \gamma^{-1} = \gamma\alpha\tilde{f}_*(\gamma^{-1}) \circ \tilde{f}$.

Theorem 3 ([51]) *Lifting classes of f are in 1-1 correspondence with $\tilde{f}_*$-conjugacy classes in π, the lifting class $[\alpha \circ \tilde{f}]$ corresponds to the $\tilde{f}_*$-cojugacy class of α.*

By an abuse of language, we will say that the fixed point class $p(\text{Fix} (\alpha \circ \tilde{f}))$, which is labeled with the lifting class $[\alpha \circ \tilde{f}]$, corresponds to the $\tilde{f}_*$-conjugacy class of α. Thus the $\tilde{f}_*$-conjugacy classes in π serve as coordinates for the fixed point classes of f, once a reference lifting $\tilde{f}$ is chosen.

A reasonable approach to finding a lower bounds for the Reidemeister number, is to consider a homomorphisms from π sending an $\tilde{f}_*$-conjugacy class to one element:

Lemma 6 ([51]) *The composition* $\eta \circ \theta$,

$$\pi = \pi_1(X, x_0) \xrightarrow{\theta} H_1(X) \xrightarrow{\eta} \mathrm{Coker} \left[H_1(X) \xrightarrow{1-\tilde{f}_{1*}} H_1(X) \right],$$

where θ is abelianization and η is the natural projection, sends every $\tilde{f}_*$-conjugacy class to a single element. Moreover, any group homomorphism $\zeta : \pi \to G$ which sends every $\tilde{f}_*$-conjugacy class to a single element, factors through $\eta \circ \theta$.

The first part of this lemma is trivial. If $\alpha' = \gamma \alpha \tilde{f}_*(\gamma^{-1})$, then

$$\theta(\alpha') = \theta(\gamma) + \theta(\alpha) + \theta(\tilde{f}_*(\gamma^{-1})) =$$

$$= \theta(\gamma) + \theta(\alpha) - f_{1*}(\theta(\gamma)) = \theta(\alpha) + (1 - f_{1*})\theta(\gamma),$$

hence $\eta \circ \theta(\alpha) = \eta \circ \theta(\alpha')$

This lemma shows the importance of the group Coker $(1 - f_{1*})$. For example

$$R(f) \geq \#\mathrm{Coker}\ (1 - f_{1*}).$$

Definition 5 *A map $f : X \to X$ is said to be eventually commutative if there exists an natural number n such that $f_*^n(\pi_1(X, x_0))$ $(\subset \pi_1(X, f^n(x_0)))$ is commutative.*

By means of Lemma 4, it is easily seen that f is eventually commutative iff $\tilde{f}_*$ is eventually commutative (see [51])

Theorem 4 ([51]) *If f is eventually commutative , then*

$$R(f) = \#\mathrm{Coker}\ (1 - f_{1*}).$$

1.3.2   Reidemeister numbers of a group endomorphism

Let G be a group and $\phi : G \to G$ an endomorphism. Two elements $\alpha, \alpha' \in G$ are said to be $\phi-conjugate$ iff there exists $\gamma \in G$ such that $\alpha' = \gamma \cdot \alpha \cdot \phi(\gamma)^{-1}$. The number of ϕ-conjugacy classes is called the *Reidemeister number* of ϕ, denoted by $R(\phi)$. We shall write $\{g\}$ for the ϕ-conjugacy class of an element $g \in G$. We shall also write $\mathcal{R}(\phi)$ for the set of ϕ-conjugacy classes of elements of G. If ϕ is the identity map then the ϕ-conjugacy classes are the usual conjugacy classes in the group G.

An endomorphism $\phi : G \to G$ is said to be eventually commutative if there exists a natural number n such that the subgroup $\phi^n(G)$ is commutative.

We are now ready to compare the Reidemeister number of an endomorphism ϕ with the Reidemeister number of $H_1(\phi) : H_1(G) \to H_1(G)$, where $H_1 = H_1^{Gp}$ is the first integral homology functor from groups to abelian groups.

Theorem 5 ([51]) *If $\phi : G \to G$ is eventually commutative, then*

$$R(\phi) = R(H_1(\phi)) = \#\text{Coker}\,(1 - H_1(\phi)).$$

This means that to find out about the Reidemeister numbers of eventually commutative endomorphisms, it is sufficient to study the Reidemeister numbers of endomorphisms of abelian groups.

From theorem 3 it follows

Corollary 1 $R(f) = R(\tilde{f}_*)$

So we see that the homotopy invariant - the Reidemeister number of a continuous map is the same as the Reidemeister number of the induced endomorphism $\tilde{f}_*$ on the fundamental group. The following are simple but very useful facts about ϕ-conjugacy classes.

Lemma 7 ([51]) *If G is a group and ϕ is an endomorphism of G then an element $x \in G$ is always ϕ-conjugate to its image $\phi(x)$.*

PROOF If $g = x^{-1}$ then one has immediately $gx = \phi(x)\phi(g)$. The existence of a g satisfying this equation implies that x and $\phi(x)$ are ϕ-conjugate.

1.4   Nielsen numbers of a continuous map

1.4.1   The fixed point index

The fixed point index provides an algebraic count of fixed points. We will introduce step-by-step construction of the index, and list without proof the most useful properties. The reader may consult the book of Dold .

(A) THE INDEX OF AN ISOLATED FIXED POINT IN R^n.

Suppose $R^n \supset U \xrightarrow{f} R^n$, and $a \in U$ is an isolated fixed point of f. Pick a sphere S_a^{n-1} centered at a, small enough to exclude other fixed points. On S_a^{n-1}, the vector $x - f(x) \neq 0$, so a direction field

$$\phi : S_a^{n-1} \to S^{n-1}, \; \phi(x) = \frac{x - f(x)}{|x - f(x)|},$$

is defined.

Definition 6 Index *(f,a)= degree of ϕ*

Example 4 *If f is a constant map to the point a, then* Index $(f, a) = 1$

Example 5 *Suppose f is differentiable at a with Jacobian $A = (\frac{\partial f}{\partial x})_a$, and* $\det(I - A) \neq 0$. *Then a is an isolated fixed point and*

$$\text{Index } (f, a) = \text{sign } \det(I - A) = (-1)^k,$$

where k is the number (counted with multiplicity) of real eigenvalues of A greater than 1.

(B) FIXED POINT INDEX IN R^n.

Definition 7 *Suppose $R^n \supset U \xrightarrow{f} R^n$, and $Fix(f)$ is compact. Take any open set $V \subset U$ such that* Fix $(f) \subset V \subset \bar{V} \subset U$ *and $\bar{V}$ is a smooth n-manifold, then $\phi(\partial V) = i \cdot S^{n-1}$ in the homological sense for some $i \in \mathbb{Z}$. This i is independent of the choice of V, and is defined to be the fixed point index of f on U, denoted* Index (f, U).

Another way of defining Index (f, U) is by approximation, namely, approximate f by a smooth map with only generic fixed points(isolated fixed points satisfying the condition in example 5), and add up their indices.

(C) FIXED POINT INDEX FOR POLYHEDRA.

Every compact polyhedron can be embedded in some Euclidean space as a neighborhood retract. Suppose now we are given a compact polyhedron X, and a map $X \supset U \xrightarrow{f} X$. X can be imbedded in R^N with inclusion

$X \xrightarrow{i} R^N$. And there is a neighborhood W of $i(X)$ in R^N and a retraction $W \xrightarrow{r} X$ such that $r \circ i = id_X$. We have a diagram

$$
\begin{array}{ccccc}
X & \supset & U & \xrightarrow{f} & X \\
\uparrow r & & \uparrow r & & \downarrow i \\
R^N \supset W & \supset & r^{-1}(U) & \xrightarrow{i \circ f \circ r} & R^N
\end{array} .
$$

Definition 8 *When $Fix(f)$ is compact, define the fixed point index to be*

$$\text{Index } (f, U) := \text{Index } (i \circ f \circ r, r^{-1}(U)),$$

the later being the index in R^N. It is independent of the choice of N, W, i and r.

All the facts we need about the fixed point index are listed below.

(I) Existence of fixed points. If Index $(f, U) \neq 0$, then f has at least one fixed point in U.

(II) Homotopy invariance. If $H = \{h_t\} : f_0 \simeq f_1 : U \to X$ is a homotopy such that $\cup_{t \in I} \text{Fix } (h_t)$ is compact, then

$$\text{Index } (f_0, U) = \text{Index } (f_1, U).$$

(III) Additivity. Suppose $U_1, ..., U_s$ are disjoint open subsets of U, and f has no fixed points on $U - \cup_{j=1}^{s} U_j$. If Index (f, U) is defined, then Index (f, U_j), $j = 1, ..., s$ are all defined and

$$\text{Index } (f, U) = \sum_{j=1}^{s} \text{Index } (f, U_j).$$

(IV) Normalization. If $f : X \to X$, then

$$\text{Index } (f, X) = L(f) := \sum_{k=0}^{\dim X} (-1)^k \text{Tr} \left[f_{*k} : H_k(X; \mathbb{Q}) \to H_k(X; \mathbb{Q}) \right]$$

where $L(f)$ is the Lefschetz number of f.

Lemma 8 ([51]) *(Homotopy invariance) Let X be a connected, compact polyhedron, $H = \{h_t\} : f_0 \simeq f_1 : X \to X$ be a homotopy, and F_i be a fixed point class of $f_i, i = 0, 1$. If F_0 corresponds to F_1 via H, then*

$$\text{Index } (f_0, F_0) = \text{Index } (f_1, F_1)$$

1.4.2   Nielsen numbers

Definition 9 *A fixed point class is called essential if its index is nonzero. The number of essential fixed point classes is called the Nielsen number of f, denoted by $N(f)$.*

The next lemma follows directly from the definitions and the properties of the index.

Lemma 9 *(1) $N(f) \leq R(f)$.*
   (2) Each essential fixed point class is non-empty.
   (3) $N(f)$ is non-negative integer, $0 \leq N(f) < \infty$.
   (4) $N(f) \leq \#\mathrm{Fix}\ (f)$.
   (5) The sum of the indices of all essential fixed point classes of f equals to the Lefschetz number $L(f)$. Hence $L(f) \neq 0$ implies $N(f) \geq 1$.

Example 6

$$N(id_X) = \begin{cases} 1 & \text{if Euler characteristic } \chi(X) \neq 0, \\ 0 & \text{if } \chi(X) = 0 \end{cases}$$

Lemma 10 *$N(f)$ is a homotopy invariant of f. Every map homotopic to f has at least $N(f)$ fixed points. In other words, $N(f) \leq Min\{\#Fix(g)|g \simeq f\}$.*

It is this lemma that made the Nielsen number so important in fixed point theory. Lemma tells us that an essential fixed point class can never disappear (i.e. become empty) via a homotopy.

   Two maps $f : X \to X$ and $g : Y \to Y$ are said to be of the same homotopy type if there is a homotopy equivalence $h : X \to Y$ such that $h \circ f \simeq g \circ h$.

Theorem 6 ([51]) *(Homotopy type invariance of the Nielsen number). If X, Y are compact connected polyhedra, and $f : X \to X$ and $g : Y \to Y$ are of the same homotopy type, then $N(f) = N(g)$.*

1.4.3   The least number of fixed points

Let X be a compact connected polyhedron , and let $f : X \to X$ be a map. Consider the number

$$MF[f] := Min\{\#\text{Fix }(g)|g \simeq f\},$$

i.e. the least number of fixed points in the homotopy class $[f]$ of f. We know that $N(f)$ is a lower bound for $MF[f]$. The importance of the Nielsen number in the fixed point theory lies in the fact that, under a mild restriction on the space involved , it is indeed the minimal number of fixed points in the homotopy class.The equality $MF[f] = N(f)$ means that we can homotope the map f so that each essential fixed point class is combined into a single fixed point and each inessential fixed point class is removed.

Definition 10 *A point x of a connected space X is a (global) separating point of X if $X - x$ is not connected. A point x of a space X is a local separating point if x is a separating point of some connected open subspace U of X.*

Theorem 7 ([51]) *Let X be a compact connected polyhedron without local separating points. Suppose X is not a surface (closed or with boundary) of negative Euler characteristic. Then $MF[f] = N(f)$ for any map $f : X \to X$.*

This theorem is generalization of the classical theorem of Wecken [97] for manifolds of dimension ≥ 3.

Lemma 11 ([51]) *(Geometric characterization of the Nielsen number). In the category of compact connected polyhedra, the Nielsen number of a self-map equals the least number of fixed points among all self-maps having the same homotopy type.*

PROOF Let $f : X \to X$ be a self -map in the category , and consider the number $m = Min\{\#\text{Fix }(g)|g \text{ has the same homotopy type as } f\}$.

Then $N(f) \leq m$ by lemma 9 and theorem 6 . On the other hand , there always exists a manifold M (with boundary) of dimension ≥ 3 which has the same homotopy type as X (for example , $M=$ the regular neighborhood of X imbedded in a Euclidean space). So , by the theorem 7, the lower bound $N(f)$ is realizable on M by a map having the same homotopy type as f.

Chapter 2

The Reidemeister zeta function

PROBLEM. For which groups and endomorphisms is the Reidemeister zeta function a rational function? When does it have a functional equation? Is $R_\phi(z)$ an algebraic function?

2.1   A Convolution Product

When $R_\phi(z)$ is a rational function the infinite sequence $\{R(\phi^n)\}_{n=1}^\infty$ of Reidemeister numbers is determined by a finite set of complex numbers - the zeros and poles of $R_\phi(z)$.

Lemma 12 $R_\phi(z)$ *is a rational function if and only if there exists a finite set of complex numbers α_i and β_j such that $R(\phi^n) = \sum_j \beta_j^n - \sum_i \alpha_i^n$ for every $n > 0$.*

PROOF Suppose $R_\phi(z)$ is a rational function. Then

$$R_\phi(z) = \frac{\prod_i (1 - \alpha_i z)}{\prod_j (1 - \beta_j z)},$$

where $\alpha_i, \beta_j \in \mathbb{C}$. Taking the logarithmic derivative of both sides and then using the geometric series expansion we see that $R(\phi^n) = \sum_j \beta_j^n - \sum_i \alpha_i^n$. The converse is proved by a direct calculation.

For two sequences (x_n) and (y_n) we may define the corresponding zeta functions:

$$X(z) := \exp\left(\sum_{n=1}^\infty \frac{x_n}{n} z^n\right),$$

27

$$Y(z) := \exp\left(\sum_{n=1}^{\infty} \frac{y_n}{n} z^n\right).$$

Alternately, given complex functions X and Y (defined in a neighborhood of 0) we may define sequences

$$x_n := \frac{d^n}{dz^n} \log\left(X(z)\right)\big|_{z=0},$$

$$y_n := \frac{d^n}{dz^n} \log\left(Y(z)\right)\big|_{z=0}.$$

Taking the componentwise product of the two sequences gives another sequence, from which we obtain another complex function. We call this new function the *additive convolution* of X and Y, and we write it

$$(X * Y)(z) := \exp\left(\sum_{n=1}^{\infty} \frac{x_n \cdot y_n}{n} z^n\right).$$

It follows immediately from lemma 1 that if X and Y are rational functions then $X * Y$ is a rational function. In fact we may show using the same method the following

Lemma 13 (Convolution of rational functions) *Let*

$$X(z) = \prod_i (1 - \alpha_i z)^{m(i)}, \quad Y(z) = \prod_j (1 - \beta_j z)^{l(j)}$$

be rational functions in z. Then $X * Y$ is the following rational function

$$(X * Y)(z) = \prod_{i,j} (1 - \alpha_i \beta_j z)^{-m(i).l(j)}. \tag{2.1}$$

A consequence of this is the following

Lemma 14 (Functional equation of a convolution) *Let $X(z)$ and $Y(z)$ be rational functions satisfying the following functional equations*

$$X\left(\frac{1}{d_1.z}\right) = K_1 z^{-e_1} X(z)^{f_1}, \quad Y\left(\frac{1}{d_2.z}\right) = K_2 z^{-e_2} Y(z)^{f_2},$$

with $d_i \in \mathbb{C}^\times$, $e_i \in \mathbb{Z}$, $K_i \in \mathbb{C}^\times$ and $f_i \in \{1, -1\}$. *Suppose also that* $X(0) = Y(0) = 1$. *Then the rational function* $X * Y$ *has the following functional equation:*

$$(X * Y)\left(\frac{1}{d_1 d_2 z}\right) = K_3 z^{-e_1 e_2}(X * Y)(z)^{f_1 f_2} \tag{2.2}$$

for some $K_3 \in \mathbb{C}^\times$.

PROOF The functions X and Y have representations of the following form:

$$X(z) = \prod_{i=1}^{a}(1 - \alpha_i z)^{m(i)}, \quad Y(z) = \prod_{j=1}^{b}(1 - \beta_j z)^{l(j)}.$$

The functional equation for X means that if $z_0 \in \mathbb{C}^\times$ is a zero or pole of X with multiplicity m, then $\frac{1}{d_1 . z_0}$ must be a pole or zero of X with multiplicity $f_1 m$. We therefore have a map

$$i \longmapsto i'$$

defined by the equation

$$\alpha_i = \frac{1}{d_1 \alpha_{i'}}$$

and with the property that

$$m(i) = f_1 m(i').$$

Similarly there is a map $j \mapsto j'$ with corresponding properties:

$$\beta_j = \frac{1}{d_2 \beta_{j'}},$$

$$l(j) = f_2 l(j').$$

Putting these maps together, we obtain a map of pairs

$$(i, j) \longmapsto (i', j')$$

with the following properties

$$\alpha_i \beta_j = \frac{1}{d_1 d_2 \alpha_{i'} \beta_{j'}},$$

$$m(i)l(j) = f_1 f_2 m(i')l(j').$$

It follows from the previous lemma that the functions $(X * Y)(z)$ and $(X * Y)(\frac{1}{d_1 d_2 z})^{f_1 f_2}$ have the same zeros and poles in the domain $\mathbb{C}^\times$. Since both functions are rational, we deduce that $X * Y$ has a functional equation of the form

$$(X * Y)\left(\frac{1}{d_1 d_2 z}\right) = K_3 z^{-e_3} (X * Y)(z)^{f_1 f_2},$$

for some $e_3 \in \mathbb{Z}$. It remains only to show that $e_3 = e_1 e_2$.

By comparing the degrees at zero of the functions $X(z)$ and $X(\frac{1}{d_1 z})$ we obtain from the functional equation for X:

$$e_1 = \sum_{i=1}^{a} m(i).$$

Similarly we have

$$e_2 = \sum_{j=1}^{b} l(j).$$

For the same reasons we have

$$
\begin{aligned}
e_3 &= \sum_{i,j} m(i)l(j) \\
&= \left(\sum_{i=1}^{a} m(i)\right)\left(\sum_{j=1}^{b} l(j)\right) \\
&= e_1 e_2.
\end{aligned}
$$

2.2   Pontryagin Duality

Let G be a locally compact Abelian topological group. We write $\hat{G}$ for the set of continuous homomorphisms from G to the circle $U(1) = \{z \in \mathbb{C} : |z| = 1\}$. This is a group with pointwise multiplication. We call $\hat{G}$ the *Pontryagin dual* of G. When we equip $\hat{G}$ with the compact-open topology it becomes a locally compact Abelian topological group. The dual of the dual of G is canonically isomorphic to G.

A continuous endomorphism $f : G \to G$ gives rise to a continuous endomorphism $\hat{f} : \hat{G} \to \hat{G}$ defined by

$$\hat{f}(\chi) := \chi \circ f.$$

There is a 1-1 correspondence between the closed subgroups H of G and the quotient groups $\hat{G}/H^*$ of $\hat{G}$ for which H^* is closed in $\hat{G}$. This correspondence is given by the following:

$$H \leftrightarrow \hat{G}/H^*,$$

$$H^* := \{\chi \in \hat{G} \mid H \subset \ker \chi\}.$$

Under this correspondence, $\hat{G}/H^*$ is canonically isomorphic to the Pontryagin dual of H. If we identify G canonically with the dual of $\hat{G}$ then we have $H^{**} = H$.

If G is a finitely generated free Abelian group then a homomorphism $\chi : G \to U(1)$ is completely determined by its values on a basis of G, and these values may be chosen arbitrarily. The dual of G is thus a torus whose dimension is equal to the rank of G.

If $G = \mathbb{Z}/n\mathbb{Z}$ then the elements of $\hat{G}$ are of the form

$$x \to e^{\frac{2\pi i y x}{n}}$$

with $y \in \{1, 2, \ldots, n\}$. A cyclic group is therefore (uncanonically) isomorphic to itself.

The dual of $G_1 \oplus G_2$ is canonically isomorphic to $\hat{G}_1 \oplus \hat{G}_2$. From this we see that any finite abelian group is (non-canonically) isomorphic to its own Pontryagin dual group, and that the dual of any finitely generated discrete Abelian group is the direct sum of a Torus and a finite group.

Proofs of all these statements may be found, for example in [81]. We shall require the following statement:

Proposition 2 *Let $\phi : G \to G$ be an endomorphism of an Abelian group G. Then the kernel $\ker \left[\hat{\phi} : \hat{G} \to \hat{G} \right]$ is canonically isomorphic to the Pontryagin dual of $\operatorname{Coker} \phi$.*

PROOF We construct the isomorphism explicitly. Let χ be in the dual of $\operatorname{Coker} (\phi : G \to G)$. In that case χ is a homomorphism

$$\chi : G/\operatorname{Im} (\phi) \longrightarrow U(1).$$

There is therefore an induced map

$$\overline{\chi} : G \longrightarrow U(1)$$

which is trivial on Im (ϕ). This means that $\overline{\chi} \circ \phi$ is trivial, or in other words $\hat{\phi}(\overline{\chi})$ is the identity element of $\hat{G}$. We therefore have $\overline{\chi} \in \ker(\hat{\phi})$.

If on the other hand we begin with $\overline{\chi} \in \ker(\hat{\phi})$, then it follows that χ is trivial on Im ϕ, and so $\overline{\chi}$ induces a homomorphism

$$\chi : G/\text{Im } (\phi) \longrightarrow U(1)$$

and χ is then in the dual of Coker ϕ. The correspondence $\chi \leftrightarrow \overline{\chi}$ is clearly a bijection.

2.3   Eventually commutative endomorphisms

An endomorphism $\phi : G \to G$ is said to be eventually commutative if there exists a natural number n such that the subgroup $\phi^n(G)$ is commutative.

2.3.1   Trace formula for the Reidemeister numbers of eventually commutative endomorphisms

We know from theorem 5 in Chapter 1 that if $\phi : G \to G$ is eventually commutative, then

$$R(\phi) = R(H_1(\phi)) = \#\text{Coker } (1 - H_1(\phi))$$

This means that to find out about the Reidemeister numbers of eventually commutative endomorphisms, it is sufficient to study the Reidemeister numbers of endomorphisms of abelian groups. For the rest of this section G will be a finitely generated abelian group.

Lemma 15 *Let $\phi : \mathbb{Z}^k \to \mathbb{Z}^k$ be a group endomorphism. Then we have*

$$R(\phi) = (-1)^{r+p} \sum_{i=0}^{k} (-1)^i \text{Tr } (\Lambda^i \phi). \tag{2.3}$$

where p the number of $\mu \in \text{Spec } \phi$ such that $\mu < -1$, and r the number of real eigenvalues of ϕ whose absolute value is > 1. Λ^i denotes the exterior power.

PROOF Since $\mathbb{Z}^k$ is Abelian, we have as before,

$$R(\phi) = \#\text{Coker } (1 - \phi).$$

On the other hand we have

$$\#\text{Coker } (1 - \phi) = |\det(1 - \phi)|,$$

and hence $R(\phi) = (-1)^{r+p} \det(1 - \phi)$ (complex eigenvalues contribute nothing to the signdet$(1 - \phi)$ since they come in conjugate pairs and $(1 - \lambda)(1 - \bar{\lambda}) = |1 - \lambda|^2 > 0$). It is well known from linear algebra that $\det(1 - \phi) = \sum_{i=0}^{k}(-1)^i \text{Tr } (\Lambda^i \phi)$. From this we have the trace formula for Reidemeister number.

Now let ϕ be an endomorphism of finite Abelian group G.Let V be the complex vector space of complex valued functions on the group G.The map ϕ induces a linear map $A : V \rightarrow V$ defined by

$$A(f) := f \circ \phi.$$

Lemma 16 *Let $\phi : G \rightarrow G$ be an endomorphism of a finite Abelian group G. Then we have*

$$R(\phi) = TrA \qquad (2.4)$$

We give two proofs of this lemma . The first proof is given here and the second proof is a special case of the proof of theorem 15

PROOF The characteristic functions of the elements of G form a basis of V,and are mapped to one another by A(the map need not be a bijection).Therefore the trace of A is the number of elements of this basis which are fixed by A.On the other hand, since G is Abelian, we have,

$$
\begin{aligned}
R(\phi) &= \#\text{Coker } (1 - \phi) \\
&= \#G/\#\text{Im } (1 - \phi) \\
&= \#G/\#(G/\ker(1 - \phi)) \\
&= \#G/(\#G/\#\ker(1 - \phi)) \\
&= \#\ker(1 - \phi) \\
&= \#\text{Fix } (\phi)
\end{aligned}
$$

We therefore have $R(\phi) = \#\text{Fix } (\phi) = \text{Tr } A$.

For a finitely generated Abelian group G we define the finite subgroup G^{finite} to be the subgroup of torsion elements of G. We denote the quotient $G^\infty := G/G^{finite}$. The group G^∞ is torsion free. Since the image of any torsion element by a homomorphism must be a torsion element, the endomorphism $\phi : G \to G$ induces endomorphisms

$$\phi^{finite} : G^{finite} \longrightarrow G^{finite}, \quad \phi^\infty : G^\infty \longrightarrow G^\infty.$$

As above, the map ϕ^{finite} induces a linear map $A : V \to V$, where V be the complex vector space of complex valued functions on the group G^{finite}.

Theorem 8 *If G is a finitely generated Abelian group and ϕ an endomorphism of G. Then we have*

$$R(\phi) = (-1)^{r+p} \sum_{i=0}^{k} (-1)^i \mathrm{Tr}\, (\Lambda^i \phi^\infty \otimes A). \tag{2.5}$$

where k is rgG^∞, p the number of $\mu \in \mathrm{Spec}\, \phi^\infty$ such that $\mu < -1$, and r the number of real eigenvalues of ϕ^∞ whose absolute value is > 1.

PROOF By proposition 2, the cokernel of $(1-\phi) : G \to G$ is the Pontrjagin dual of the kernel of the dual map $\widehat{(1-\phi)} : \hat{G} \to \hat{G}$. Since Coker $(1-\phi)$ is finite, we have

$$\#\mathrm{Coker}\, (1-\phi) = \#\ker (\widehat{1-\phi}).$$

The map $\widehat{1-\phi}$ is equal to $\hat{1} - \hat{\phi}$. Its kernel is thus the set of fixed points of the map $\hat{\phi} : \hat{G} \to \hat{G}$. We therefore have

$$R(\phi) = \#\mathrm{Fix}\, \left(\hat{\phi} : \hat{G} \to \hat{G} \right) \tag{2.6}$$

The dual group of G^∞ is a torus whose dimension is the rank of G. This is canonically a closed subgroup of $\hat{G}$. We shall denote it $\hat{G}_0$. The quotient $\hat{G}/\hat{G}_0$ is canonically isomorphic to the dual of G^{finite}. It is therefore finite. From this we know that $\hat{G}$ is a union of finitely many disjoint tori. We shall call these tori $\hat{G}_0, \ldots, \hat{G}_t$.

We shall call a torus $\hat{G}_i$ periodic if there is an iteration $\hat{\phi}^s$ such that $\hat{\phi}^s(\hat{G}_i) \subset \hat{G}_i$. If this is the case, then the map $\hat{\phi}^s : \hat{G}_i \to \hat{G}_i$ is a translation of the map $\hat{\phi}^s : \hat{G}_0 \to \hat{G}_0$ and has the same number of fixed points as this map. If $\hat{\phi}^s(\hat{G}_i) \not\subset \hat{G}_i$ then $\hat{\phi}^s$ has no fixed points in $\hat{G}_i$. From this we see

$$\#\mathrm{Fix}\left(\hat{\phi} : \hat{G} \to \hat{G}\right) = \#\mathrm{Fix}\left(\hat{\phi} : \hat{G}_0 \to \hat{G}_0\right) \times \#\{\hat{G}_i \mid \hat{\phi}(\hat{G}_i) \subset \hat{G}_i\}.$$

We now rephrase this

$$\begin{aligned} &\#\mathrm{Fix}\left(\hat{\phi} : \hat{G} \to \hat{G}\right) \\ &= \#\mathrm{Fix}\left(\widehat{\phi^\infty} : \hat{G}_0 \to \hat{G}_0\right) \times \#\mathrm{Fix}\left(\widehat{\phi^{finite}} : \hat{G}/(\hat{G}_0) \to \hat{G}/(\hat{G}_0)\right). \end{aligned}$$

From this we have product formula for Reidemeister numbers

$$R(\phi) = R(\phi^\infty) \cdot R(\phi^{finite}).$$

The trace formula for $R(\phi)$ follow from the previous two lemmas and formula

$$\mathrm{Tr}\left(\Lambda^i \phi^\infty\right) \cdot \mathrm{Tr}\left(A\right) = \mathrm{Tr}\left(\Lambda^i \phi^\infty \otimes A\right).$$

2.3.2 Rationality of Reidemeister zeta functions of eventually commutative endomorphisms - first proof.

If we compare the Reidemeister zeta function of an endomorphism ϕ with the Reidemeister zeta function of $H_1(\phi) : H_1(G) \to H_1(G)$, where $H_1 = H_1^{Gp}$ is the first integral homology functor from groups to abelian groups, then we have from theorem 5 in chapter 1 following

Theorem 9 *If $\phi : G \to G$ is eventually commutative, then*

$$R_\phi(z) = R_{H_1(\phi)}(z) = \exp\left(\sum_{n=1}^\infty \frac{\#\mathrm{Coker}\left(1 - H_1(\phi)^n\right)}{n} z^n\right).$$

This means that to find out about the Reidemeister zeta functions of eventually commutative endomorphisms, it is sufficient to study the zeta functions of endomorphisms of Abelian groups.

Lemma 17 *Let $\phi : \mathbb{Z}^k \to \mathbb{Z}^k$ be a group endomorphism. Then we have*

$$R_\phi(z) = \left(\prod_{i=0}^k \det(1 - \Lambda^i \phi \cdot \sigma \cdot z)^{(-1)^{i+1}}\right)^{(-1)^r} \tag{2.7}$$

where $\sigma = (-1)^p$ with p the number of $\mu \in \mathrm{Spec}\ \phi$ such that $\mu < -1$, and r the number of real eigenvalues of ϕ whose absolute value is > 1. Λ^i denotes the exterior power.

PROOF Since $\mathbb{Z}^k$ is Abelian, we have as before,

$$R(\phi^n) = \#\text{Coker } (1 - \phi^n).$$

On the other hand we have

$$\#\text{Coker } (1 - \phi^n) = |\det(1 - \phi^n)|,$$

and hence $R(\phi^n) = (-1)^{r+pn} \det(1-\phi^n)$. It is well known from linear algebra that $\det(1-\phi^n) = \sum_{i=0}^{k}(-1)^i \text{Tr } (\Lambda^i \phi^n)$. From this we have the following trace formula for Reidemeister numbers:

$$R(\phi^n) = (-1)^{r+pn} \sum_{i=0}^{k}(-1)^i \text{Tr } (\Lambda^i \phi^n). \tag{2.8}$$

We now calculate directly

$$
\begin{aligned}
R_\phi(z) &= \exp\left(\sum_{n=1}^{\infty} \frac{R(\phi^n)}{n} z^n \right) \\
&= \exp\left(\sum_{n=1}^{\infty} \frac{(-1)^r \sum_{i=0}^{k}(-1)^i \text{Tr } (\Lambda^i \phi^n)}{n} (\sigma z)^n \right) \\
&= \left(\prod_{i=0}^{k} \left(\exp\left(\sum_{n=1}^{\infty} \frac{1}{n} \text{Tr } (\Lambda^i \phi^n) \cdot (\sigma z)^n \right) \right)^{(-1)^i} \right)^{(-1)^r} \\
&= \left(\prod_{i=0}^{k} \det\left(1 - \Lambda^i \phi \cdot \sigma z \right)^{(-1)^{i+1}} \right)^{(-1)^r}.
\end{aligned}
$$

Lemma 18 *Let $\phi : G \to G$ be an endomorphism of a finite Abelian group G. Then we have*

$$R_\phi(z) = \prod_{[\gamma]} \frac{1}{1 - z^{\#[\gamma]}} \tag{2.9}$$

where the product is taken over the periodic orbits of ϕ in G.

We give two proofs of this lemma. The first proof is given here and the second proof is a special case of the proof of theorem 16 .

PROOF Since G is Abelian, we again have,

$$
\begin{aligned}
R(\phi^n) &= \#\mathrm{Coker}\ (1 - \phi^n) \\
&= \#G / \#\mathrm{Im}\ (1 - \phi^n) \\
&= \#G / \#(G / \ker(1 - \phi^n)) \\
&= \#G / (\#G / \# \ker(1 - \phi^n)) \\
&= \# \ker(1 - \phi^n) \\
&= \#\mathrm{Fix}\ (\phi^n)
\end{aligned}
$$

We shall call an element of G periodic if it is fixed by some iteration of ϕ. A periodic element γ is fixed by ϕ^n iff n is divisible by the cardinality the orbit of γ. We therefore have

$$
\begin{aligned}
R(\phi^n) &= \sum_{\substack{\gamma\ periodic \\ \#[\gamma]\,|\,n}} 1 \\
&= \sum_{\substack{[\gamma]\ such\ that, \\ \#[\gamma]\,|\,n}} \#[\gamma].
\end{aligned}
$$

From this follows

$$
\begin{aligned}
R_\phi(z) &= \exp\left(\sum_{n=1}^{\infty} \frac{R(\phi^n)}{n} z^n \right) \\
&= \exp\left(\sum_{[\gamma]} \sum_{\substack{n=1 \\ \#[\gamma]\,|\,n}}^{\infty} \frac{\#[\gamma]}{n} z^n \right) \\
&= \prod_{[\gamma]} \exp\left(\sum_{n=1}^{\infty} \frac{\#[\gamma]}{\#[\gamma]n} z^{\#[\gamma]n} \right) \\
&= \prod_{[\gamma]} \exp\left(\sum_{n=1}^{\infty} \frac{1}{n} z^{\#[\gamma]n} \right) \\
&= \prod_{[\gamma]} \exp\left(-\log\left(1 - z^{\#[\gamma]} \right) \right) \\
&= \prod_{[\gamma]} \frac{1}{1 - z^{\#[\gamma]}}.
\end{aligned}
$$

For a finitely generated Abelian group G we define the finite subgroup G^{finite} to be the subgroup of torsion elements of G. We denote the quotient

$G^\infty := G/G^{finite}$. The group G^∞ is torsion free. Since the image of any torsion element by a homomorphism must be a torsion element, the function $\phi : G \to G$ induces maps

$$\phi^{finite} : G^{finite} \longrightarrow G^{finite}, \quad \phi^\infty : G^\infty \longrightarrow G^\infty.$$

Theorem 10 *If G is a finitely generated Abelian group and ϕ an endomorphism of G then $R_\phi(z)$ is a rational function and is equal to the following additive convolution:*

$$R_\phi(z) = R_\phi^\infty(z) * R_\phi^{finite}(z). \tag{2.10}$$

where $R_\phi^\infty(z)$ is the Reidemeister zeta function of the endomorphism $\phi^\infty :$ $G^\infty \to G^\infty$, and $R_\phi^{finite}(z)$ is the Reidemeister zeta function of the endomorphism $\phi^{finite} : G^{finite} \to G^{finite}$. The functions $R_\phi^\infty(z)$ and $R_\phi^{finite}(z)$ are given by the formulae

$$R_\phi^\infty(z) = \left(\prod_{i=0}^{k} \det(1 - \Lambda^i \phi^\infty \cdot \sigma \cdot z)^{(-1)^{i+1}} \right)^{(-1)^r}, \tag{2.11}$$

$$R_\phi^{finite}(z) = \prod_{[\gamma]} \frac{1}{1 - z^{\#[\gamma]}}. \tag{2.12}$$

with the product in (2.12) being taken over all periodic ϕ-orbits of torsion elements $\gamma \in G$. Also, $\sigma = (-1)^p$ where p is the number of real eingevalues $\lambda \in \mathrm{Spec}\, \phi^\infty$ such that $\lambda < -1$ and r is the number of real eingevalues $\lambda \in \mathrm{Spec}\, \phi^\infty$ such that $|\lambda| > 1$.

PROOF By proposition 2, the cokernel of $(1 - \phi^n) : G \to G$ is the Pontrjagin dual of the kernel of the dual map $(\widehat{1 - \phi^n}) : \hat{G} \to \hat{G}$. Since Coker $(1 - \phi^n)$ is finite, we have

$$\#\mathrm{Coker}\ (1 - \phi^n) = \#\ker(\widehat{1 - \phi^n}).$$

The map $\widehat{1 - \phi^n}$ is equal to $\hat{1} - \hat{\phi}^n$. Its kernel is thus the set of fixed points of the map $\hat{\phi}^n : \hat{G} \to \hat{G}$. We therefore have

$$R(\phi^n) = \#\mathrm{Fix}\ \left(\hat{\phi}^n : \hat{G} \to \hat{G} \right).^{[1]} \tag{2.13}$$

---

[1] We shall use this formula again later to connect the Reidemeister number of ϕ with the Lefschetz number of $\hat{\phi}$.

The dual group of G^∞ is a torus whose dimension is the rank of G. This is canonically a closed subgroup of $\hat{G}$. We shall denote it $\hat{G}_0$. The quotient $\hat{G}/\hat{G}_0$ is canonically isomorphic to the dual of G^{finite}. It is therefore finite. From this we know that $\hat{G}$ is a union of finitely many disjoint tori. We shall call these tori $\hat{G}_0, \ldots, \hat{G}_r$.

We shall call a torus $\hat{G}_i$ periodic if there is an iteration $\hat{\phi}^s$ such that $\hat{\phi}^s(\hat{G}_i) \subset \hat{G}_i$. If this is the case, then the map $\hat{\phi}^s : \hat{G}_i \to \hat{G}_i$ is a translation of the map $\hat{\phi}^s : \hat{G}_0 \to \hat{G}_0$ and has the same number of fixed points as this map. If $\hat{\phi}^s(\hat{G}_i) \not\subset \hat{G}_i$ then $\hat{\phi}^s$ has no fixed points in $\hat{G}_i$. From this we see

$$\#\text{Fix}\left(\hat{\phi}^n : \hat{G} \to \hat{G}\right) = \#\text{Fix}\left(\hat{\phi}^n : \hat{G}_0 \to \hat{G}_0\right) \times \#\{\hat{G}_i \mid \hat{\phi}^n(\hat{G}_i) \subset \hat{G}_i\}.$$

We now rephrase this

$$\#\text{Fix}\left(\hat{\phi}^n : \hat{G} \to \hat{G}\right)$$
$$= \#\text{Fix}\left(\widehat{\phi^\infty}^n : \hat{G}_0 \to \hat{G}_0\right) \times \#\text{Fix}\left(\widehat{\phi^{finite}}^n : \hat{G}/(\hat{G}_0) \to \hat{G}/(\hat{G}_0)\right).$$

From this we have that $R(\phi^n) = R((\phi^\infty)^n) \cdot R((\phi^{finite})^n)$ for every n and

$$R_\phi(z) = R_{(\phi^\infty)}(z) * R_{(\phi^{finite})}(z).$$

The rationality of $R_\phi(z)$ and the formulae for $R_\phi^\infty(z)$ and $R_\phi^{finite}(z)$ follow from the previous two lemmas and lemma 13.

Corollary 2 *Let the assumptions of theorem 10 hold. Then the poles and zeros of the Reidemeister zeta function are complex numbers of the form $\zeta^a b$ where b is the reciprocal of an eigenvalue of one of the matrices*

$$\Lambda^i(\phi^\infty) : \Lambda^i(G^\infty) \longrightarrow \Lambda^i(G^\infty) \qquad 0 \leq i \leq \text{rank } G$$

and ζ^a is a ψth root of unity where ψ is the number of periodic torsion elements in G. The multiplicities of the roots or poles $\zeta^a b$ and $\zeta^{a'} b'$ are the same if $b = b'$ and $hcf(a, \psi) = hcf(a', \psi)$.

2.3.3 Functional equation for the Reidemeister zeta function of an eventually commutative endomorphism

Lemma 19 (Functional equation for the torsion free part)

Let $\phi : \mathbb{Z}^k \to \mathbb{Z}^k$ be an endomorphism. The Reidemeister zeta function $R_\phi(z)$ has the following functional equation:

$$R_\phi\left(\frac{1}{dz}\right) = \epsilon_1 \cdot R_\phi(z)^{(-1)^k}. \tag{2.14}$$

where $d = \det \phi$ and ϵ_1 are a constants in $\mathbb{C}^\times$.

PROOF Via the natural nonsingular pairing $(\Lambda^i \mathbb{Z}^k) \otimes (\Lambda^{k-i} \mathbb{Z}^k) \to \mathbb{C}$ the operators $\Lambda^{k-i}\phi$ and $d \cdot (\Lambda^i \phi)^{-1}$ are adjoint to each other.

We consider an eigenvalue λ of $\Lambda^i \phi$. By lemma 17, this contributes a term $\left((1 - \frac{\lambda\sigma}{dz})^{(-1)^{i+1}}\right)^{(-1)^r}$ to $R_\phi\left(\frac{1}{dz}\right)$. We rewrite this term as

$$\left(\left(1 - \frac{d\sigma z}{\lambda}\right)^{(-1)^{i+1}} \left(\frac{-dz}{\lambda\sigma}\right)^{(-1)^i}\right)^{(-1)^r}$$

and note that $\frac{d}{\lambda}$ is an eigenvalue of $\Lambda^{k-i}\phi$. Multiplying these terms together we obtain,

$$R_\phi\left(\frac{1}{dz}\right) = \left(\prod_{i=1}^k \prod_{\lambda^{(i)} \in \mathrm{Spec}\ \Lambda^i \phi} \left(\frac{1}{\lambda^{(i)}\sigma}\right)^{(-1)^i}\right)^{(-1)^r} \times R_\phi(z)^{(-1)^k}.$$

The variable z has disappeared because

$$\sum_{i=0}^k (-1)^i \dim \Lambda^i \mathbb{Z}^k = \sum_{i=0}^k (-1)^i \cdot C_k^i = 0.$$

Lemma 20 (Functional equation for the finite part) Let $\phi : G \to G$ be an endomorphism of a finite, Abelian group G. The Reidemeister zeta function $R_\phi(z)$ has the following functional equation:

$$R_\phi\left(\frac{1}{z}\right) = (-1)^p z^q R_\phi(z), \tag{2.15}$$

where q is the number of periodic elements of ϕ in G and p is the number of periodic orbits of ϕ in G.

PROOF This is a simple calculation. We begin with formula (2.9).

$$R_\phi\left(\frac{1}{z}\right) = \prod_{[\gamma]} \frac{1}{1 - z^{-\#[\gamma]}}$$

$$= \prod_{[\gamma]} \frac{z^{\#[\gamma]}}{z^{\#[\gamma]} - 1}$$

$$= \prod_{[\gamma]} \frac{-z^{\#[\gamma]}}{1 - z^{\#[\gamma]}}$$

$$= \prod_{[\gamma]} -z^{\#[\gamma]} \times \prod_{[\gamma]} \frac{1}{1 - z^{\#[\gamma]}}$$

$$= \prod_{[\gamma]} -z^{\#[\gamma]} \times R_\phi(z).$$

The statement now follows because $\sum_{[\gamma]} \#[\gamma] = q$.

Theorem 11 (Functional equation) *Let $\phi : G \to G$ be an endomorphism of a finitely generated Abelian group G. If G is finite the functional equation of R_ϕ is described in lemma 20. If G is infinite then R_ϕ has the following functional equation:*

$$R_\phi\left(\frac{1}{dz}\right) = \epsilon_2 \cdot R_\phi(z)^{(-1)^{\text{Rank } G}}. \qquad (2.16)$$

where $d = \det(\phi^\infty : G^\infty \to G^\infty)$ and ϵ_2 are a constants in $\mathbb{C}^\times$.

PROOF From theorem 10 we have $R_\phi(z) = R_\phi^\infty(z) * R_\phi^{finite}(z)$. In the previous two lemmas we have obtained functional equations for the functions $R_\phi^\infty(z)$ and $R_\phi^{finite}(z)$. Lemma 14 now gives the functional equation for $R_\phi(z)$.

2.3.4   Rationality of Reidemeister zeta functions of eventually commutative endomorphisms - second proof.

Theorem 12 *Let G is a finitely generated Abelian group and ϕ an endomorphism of G .Then $R_\phi(z)$ is a rational function and is equal to*

$$R_\phi(z) = \left(\prod_{i=0}^{k} \det(1 - \Lambda^i \phi^\infty \otimes A \cdot \sigma \cdot z)^{(-1)^{i+1}}\right)^{(-1)^r} \qquad (2.17)$$

where matrix A is defined in lemma 16 , $\sigma = (-1)^p$, p , r and k are constants described in theorem 8 .

PROOF If we repeat the proof of the theorem 8 for ϕ^n instead ϕ we obtain that $R(\phi^n) = R((\phi^\infty)^n) \cdot R((\phi^{finite})^n)$. From this and lemmas 15 and 16 we have the trace formula for $R(\phi^n)$:

$$
\begin{aligned}
R(\phi^n) &= (-1)^{r+pn} \sum_{i=0}^{k} (-1)^i \operatorname{Tr} \Lambda^i (\phi^\infty)^n \cdot \operatorname{Tr} A^n \\
&= (-1)^{r+pn} \sum_{i=0}^{k} (-1)^i \operatorname{Tr} (\Lambda^i (\phi^\infty)^n \otimes A^n) \\
&= (-1)^{r+pn} \sum_{i=0}^{k} (-1)^i \operatorname{Tr} (\Lambda^i \phi^\infty \otimes A)^n.
\end{aligned}
$$

We now calculate directly

$$
\begin{aligned}
R_\phi(z) &= \exp \left(\sum_{n=1}^{\infty} \frac{R(\phi^n)}{n} z^n \right) \\
&= \exp \left(\sum_{n=1}^{\infty} \frac{(-1)^r \sum_{i=0}^{k} (-1)^i \operatorname{Tr} (\Lambda^i \phi^\infty \otimes A)^n}{n} (\sigma \cdot z)^n \right) \\
&= \left(\prod_{i=0}^{k} \left(\exp \left(\sum_{n=1}^{\infty} \frac{1}{n} \operatorname{Tr} (\Lambda^i \phi^\infty \otimes A)^n \cdot (\sigma \cdot z)^n \right) \right)^{(-1)^i} \right)^{(-1)^r} \\
&= \left(\prod_{i=0}^{k} \det \left(1 - \Lambda^i \phi^\infty \otimes A \cdot \sigma \cdot z \right)^{(-1)^{i+1}} \right)^{(-1)^r} .
\end{aligned}
$$

Corollary 3 *Let the assumptions of theorem 12 hold. Then the poles and zeros of the Reidemeister zeta function are complex numbers which are the reciprocal of an eigenvalues of one of the matrices*

$$
\Lambda^i (\phi^\infty) \otimes A \cdot \sigma \qquad 0 \le i \le \operatorname{rank} G
$$

2.3.5   Connection of the Reidemeister zeta function with the Lefschetz zeta function of the dual map

Theorem 13 (Connection with Lefschetz numbers) *Let $\phi : G \to G$ be an endomorphism of a finitely generated Abelian group. Then we have the* *following*

$$R(\phi^n) = | L(\hat{\phi}^n) |, \tag{2.18}$$

where $\hat{\phi}$ is the continuous endomorphism of $\hat{G}$ defined in section 2.2 and $L(\hat{\phi}^n)$ is the Lefschetz number of $\hat{\phi}$ thought of as a self-map of the topological space $\hat{G}$. From this it follows:

$$R_\phi(z) = L_{\hat{\phi}}(\sigma z)^{(-1)^r}, \tag{2.19}$$

where r and σ are the constants described in theorem 10. If G is finite then this reduces to

$$R(\phi^n) = L(\hat{\phi}^n) \text{ and } R_\phi(z) = L_{\hat{\phi}}(z).$$

The proof is similar to that of Anosov [4] concerning continuous maps of nil-manifolds.

PROOF We already know from formula (2.13) in the proof of theorem 10 that $R(\phi^n)$ is the number of fixed points of the map $\hat{\phi}^n$. If G is finite then $\hat{G}$ is a discrete finite set, so the number of fixed points is equal to the Lefschetz number. This finishes the proof in the case that G is finite. In general it is only necessary to check that the number of fixed points of $\hat{\phi}^n$ is equal to the absolute value of its Lefschetz number. We assume without loss of generality that $n = 1$. We are assuming that $R(\phi)$ is finite, so the fixed points of $\hat{\phi}$ form a discrete set. We therefore have

$$L(\hat{\phi}) = \sum_{x \in \text{Fix } \hat{\phi}} \text{Index } (\hat{\phi}, x).$$

Since ϕ is a group endomorphism, the zero element $0 \in \hat{G}$ is always fixed. Let x be any fixed point of $\hat{\phi}$. We then have a commutative diagram

$$
\begin{array}{ccccc}
g & \hat{G} & \xrightarrow{\hat{\phi}} & \hat{G} & g \\
\updownarrow & \updownarrow & & \updownarrow & \updownarrow \\
g+x & \hat{G} & \xrightarrow{\hat{\phi}} & \hat{G} & g+x
\end{array}
$$

in which the vertical functions are translations on $\hat{G}$ by x. Since the vertical maps map 0 to x, we deduce that

$$\text{Index } (\hat{\phi}, x) = \text{Index } (\hat{\phi}, 0)$$

and so all fixed points have the same index. It is now sufficient to show that Index $(\hat{\phi}, 0) = \pm 1$. This follows because the map on the torus

$$\hat{\phi} : \hat{G}_0 \to \hat{G}_0$$

lifts to a linear map of the universal cover, which is in this case the Lie algebra of $\hat{G}$. The index is then the sign of the determinant of the identity map minus this lifted map. This determinant cannot be zero, because $1 - \hat{\phi}$ must have finite kernel by our assumption that the Reidemeister number of ϕ is finite (if $\det(1 - \hat{\phi}) = 0$ then the kernel of $1 - \hat{\phi}$ is a positive dimensional subgroup of $\hat{G}$, and therefore infinite).

2.4   Endomorphisms of finite groups

In this section we consider finite non-Abelian groups. We shall write the group law multiplicatively. We generalize our results on endomorphisms of finite Abelian groups to endomorphisms of finite non-Abelian groups. We shall write $\{g\}$ for the ϕ-conjugacy class of an element $g \in G$. We shall write $< g >$ for the ordinary conjugacy class of g in G. We continue to write $[g]$ for the ϕ-orbit of $g \in G$, and we also write now $[< g >]$ for the ϕ-orbit of the ordinary conjugacy class of $g \in G$. We first note that if ϕ is an endomorphism of a group G then ϕ maps conjugate elements to conjugate elements. It therefore induces an endomorphism of the set of conjugacy classes of elements of G. If G is Abelian then a conjugacy class consists of a single element. The following is thus an extension of lemma 16:

Theorem 14 *Let G be a finite group and let $\phi : G \to G$ be an endomorphism. Then $R(\phi)$ is the number of ordinary conjugacy classes $< x >$ in G such that $< \phi(x) > = < x >$.*

PROOF From the definition of the Reidemeister number we have,

$$R(\phi) \;=\; \sum_{\{g\}} 1$$

where $\{g\}$ runs through the set of ϕ-conjugacy classes in G. This gives us immediately

$$R(\phi) \;=\; \sum_{\{g\}} \sum_{x \in \{g\}} \frac{1}{\#\{g\}}$$

$$= \sum_{\{g\}} \sum_{x \in \{g\}} \frac{1}{\#\{x\}}$$

$$= \sum_{x \in G} \frac{1}{\#\{x\}}.$$

We now calculate for any $x \in G$ the order of $\{x\}$. The class $\{x\}$ is the orbit of x under the G-action

$$(g, x) \longmapsto gx\phi(g)^{-1}.$$

We verify that this is actually a G-action:

$$\begin{aligned}
(id, x) &\longmapsto id.x.\phi(id)^{-1} \\
&= x, \\
(g_1 g_2, x) &\longmapsto g_1 g_2.x.\phi(g_1 g_2)^{-1} \\
&= g_1 g_2.x.(\phi(g_1)\phi(g_2))^{-1} \\
&= g_1 g_2.x.\phi(g_2)^{-1}\phi(g_1)^{-1} \\
&= g_1(g_2.x.\phi(g_2)^{-1})\phi(g_1)^{-1}.
\end{aligned}$$

We therefore have from the orbit-stabilizer theorem,

$$\#\{x\} = \frac{\#G}{\#\{g \in G \mid gx\phi(g)^{-1} = x\}}.$$

The condition $gx\phi(g)^{-1} = x$ is equivalent to

$$x^{-1}gx\phi(g)^{-1} = 1 \quad \Leftrightarrow \quad x^{-1}gx = \phi(g)$$

We therefore have

$$R(\phi) = \frac{1}{\#G} \sum_{x \in G} \#\{g \in G \mid x^{-1}gx = \phi(g)\}.$$

Changing the summation over x to summation over g, we have:

$$R(\phi) = \frac{1}{\#G} \sum_{g \in G} \#\{x \in G \mid x^{-1}gx = \phi(g)\}.$$

If $< \phi(g) >\neq< g >$ then there are no elements x such that $x^{-1}gx = \phi(g)$. We therefore have:

$$R(\phi) = \frac{1}{\#G} \sum_{\substack{g \in G \text{ such that} \\ <\phi(g)>=<g>}} \#\{x \in G \mid x^{-1}gx = \phi(g)\}.$$

The elements x such that $x^{-1}gx = \phi(g)$ form a coset of the subgroup satisfying $x^{-1}gx = g$. This subgroup is the centralizer of g in G which we write $C(g)$. With this notation we have,

$$R(\phi) = \frac{1}{\#G} \sum_{\substack{g \in G \text{ such that} \\ <\phi(g)>=<g>}} \#C(g)$$

$$= \frac{1}{\#G} \sum_{\substack{<g>\subset G \text{ such that} \\ <\phi(g)>=<g>}} \# < g > .\#C(g).$$

The last identity follows because $C(h^{-1}gh) = h^{-1}C(g)h$. From the orbit stabilizer theorem, we know that $\# < g > .\#C(g) = \#G$. We therefore have

$$R(\phi) = \#\{< g >\subset G \mid< \phi(g) >=< g >\}.$$

Let W be the complex vector space of complex valued class functions on the group G. A class function is a function which takes the same value on every element of a usual congruency class. The map ϕ induces a linear map $B : W \to W$ defined by

$$B(f) := f \circ \phi.$$

Theorem 15 *(Trace formula) Let $\phi : G \to G$ be an endomorphism of a finite group G. Then we have*

$$R(\phi) = \text{Tr } B \qquad\qquad (2.20)$$

PROOF The characteristic functions of the congruency classes in G form a basis of W, and are mapped to one another by B (the map need not be a bijection). Therefore the trace of B is the number of elements of this basis which are fixed by B. By theorem 14, this is equal to the Reidemeister number of ϕ.

From the theorem 14 we have immediately,

Theorem 16 *Let ϕ be an endomorphism of a finite group G. Then $R_\phi(z)$ is a rational function with a functional equation. In particular we have,*

$$R_\phi(z) = \prod_{[<g>]} \frac{1}{1 - z^{\#[<g>]}},$$

$$R_\phi\left(\frac{1}{z}\right) = (-1)^a z^b R_\phi(z).$$

The product here is over all periodic ϕ-orbits of ordinary conjugacy classes of elements of G. The number $\#[<g>]$ is the number of conjugacy classes in the ϕ-orbit of the conjugacy class $<g>$. In the functional equation the numbers a and b are respectively the number of periodic ϕ-orbits of conjugacy classes of elements of G and the number of periodic conjugacy classes of elements of G. A conjugacy class $<g>$ is called periodic if for some $n > 0$, $<\phi^n(g)> = <g>$

PROOF From the theorem 14 we know that $R(\phi^n)$ is the number of conjugacy classes $<g> \subset G$ such that $\phi^n(<g>) \subset <g>$. We can rewrite this

$$R(\phi^n) = \sum_{\substack{[<g>] \text{ such that} \\ \#[<g>] \,|\, n}} \#[<g>].$$

From this we have,

$$R_\phi(z) = \prod_{[<g>]} \exp\left(\sum_{\substack{n = 1 \text{ such that} \\ \#[<g>] \,|\, n}}^{\infty} \frac{\#[<g>]}{n} z^n \right).$$

The first formula now follows by using the power series expansion for $\log(1 - z)$. The functional equation follows now in exactly the same way as lemma 20 follows from lemma 18.

Corollary 4 *Suppose that ϕ_1 and ϕ_2 are two endomorphisms of a finite group G with*

$$\forall g \in G, \ \phi_1(g) = h\phi_2(g)h^{-1}$$

for some fixed element $h \in G$. Then $R_{\phi_1}(z) = R_{\phi_2}(z)$.

Corollary 5 *Let ϕ be an inner automorphism. Then*

$$R_\phi(z) = \frac{1}{(1-z)^b}$$

where b is the number of conjugacy classes of elements in the group. In particular, all but finitely many of the symmetric and alternating groups have the property that any automorphism is an inner automorphism, and so this corollary applies.

Remark 2 *If we think of the set of conjugacy classes of elements of G as a discrete set then the Reidemeister number of ϕ is equal to the Lefschetz number of the induced map on the set of the conjugacy classes of elements of G.*

Another proof of rationality of the Reidemeister zeta function for finite groups follows from the trace formula for the Reidemeister numbers in the theorem 15.

Theorem 17 *Let ϕ be an endomorphism of a finite group G. Then $R_\phi(z)$ is a rational function and given by formula*

$$R_\phi(z) = \frac{1}{\det(1 - Bz)}, \qquad (2.21)$$

Where B is defined in theorem 15

PROOF From theorem 15 it follows that $R(\phi^n) = \operatorname{Tr} B^n$ for every $n > o$. We now calculate directly

$$
\begin{aligned}
R_\phi(z) &= \exp\left(\sum_{n=1}^\infty \frac{R(\phi^n)}{n} z^n\right) = \exp\left(\sum_{n=1}^\infty \frac{\operatorname{Tr} B^n}{n} z^n\right) = \exp\left(\operatorname{Tr} \sum_{n=1}^\infty \frac{B^n}{n} z^n\right) \\
&= \exp\left(\operatorname{Tr}\left(-\log(1 - Bz)\right)\right) = \frac{1}{\det(1 - Bz)}.
\end{aligned}
$$

2.5   Endomorphisms of the direct sum of a free Abelian and a finite group

In this section let F be a finite group and k a natural number. The group G will be

$$G = \mathbb{Z}^k \oplus F$$

We shall describe the Reidemeister numbers of endomorphism $\phi : G \to G$. The torsion elements of G are exactly the elements of the finite, normal subgroup F. For this reason we have $\phi(F) \subset F$. Let $\phi^{finite} : F \to F$ be the restriction of ϕ to F, and let $\phi^\infty : G/F \to G/F$ be the induced map on the quotient group.

Let $pr_{\mathbb{Z}^k} : G \to \mathbb{Z}^k$ and $pr_F : G \to F$ be the projections onto $\mathbb{Z}^k$ and F. Then the composition

$$pr_{\mathbb{Z}^k} \circ \phi : \mathbb{Z}^k \to G \to \mathbb{Z}^k$$

is an endomorphism of $\mathbb{Z}^k$, which is given by some matrix $M \in M_k(\mathbb{Z})$. We denote by $\psi : \mathbb{Z}^k \to F$ the other component of the restriction of ϕ to $\mathbb{Z}^k$, i.e.

$$\psi(v) = pr_F(\phi(v)).$$

We therefore have for any element $(v, f) \in G$

$$\phi(v, f) = (M \cdot v, \psi(v)\phi(f)).$$

Lemma 21 *Let $g_1 = (v_1, f_1)$ and $g_2 = (v_2, f_2)$ be two elements of G. Then g_1 and g_2 are ϕ-conjugate iff*

$$v_1 \equiv v_2 \bmod (1 - M)\mathbb{Z}^k$$

and there is a $h \in F$ with

$$hf_1 = f_2\phi((1 - M)^{-1}(v_2 - v_1))\phi(h).$$

PROOF
Suppose g_1 and g_2 are ϕ-conjugate. Then there is a $g_3 = (w, h) \in G$ with $g_3 g_1 = g_2 \phi(g_3)$. Therefore

$$(w + v_1, hf_1) = (v_2 + M \cdot w, f_2\psi(w)\phi(h)).$$

Comparing the first components we obtain $(1 - M) \cdot w = v_2 - v_1$ from which it follows that v_1 is congruent to v_2 modulo $(1 - M)\mathbb{Z}^r$. Substituting $(1 - M)^{-1}(v_2 - v_1)$ for w in the second component we obtain the second relation in the lemma. The argument can easily be reversed to give the converse.

Proposition 3 *In the notation described above*

$$R(\phi) = R(\phi^{finite}) \times R(\phi^{\infty}).$$

PROOF We partition the set $\mathcal{R}(\phi)$ of ϕ-conjugacy classes of elements of G into smaller sets:

$$\mathcal{R}(\phi) = \cup_{v \in \mathbf{Z}^k/(1-M)\mathbf{Z}^k} \mathcal{R}(v)$$

where $\mathcal{R}(v)$ is the set of ϕ-conjugacy classes $\{(w,f)\}_\phi$ for which w is congruent to v modulo $(1-M)\mathbf{Z}^k$. It follows from the previous lemma that this is a partition. Now suppose $\{(w,f)\}_\phi \in \mathcal{R}(v)$. We will show that $\{(w,f)\}_\phi = \{(v,f^*)\}_\phi$ for some $f^* \in F$. This follows by setting $f^* = f\psi((1-M)^{-1}(w-v))$ and applying the previous lemma with $h = id$. Therefore $\mathcal{R}(v)$ is the set of ϕ-conjugacy classes $\{(v,f)\}_\phi$ with $f \in F$. From the previous lemma it follows that (v,f_1) and (v,f_2) are ϕ-conjugate iff there is a $h \in F$ with

$$hf_1 = f_2\psi(0)\phi(h) = f_2\phi(h)$$

This just means that f_1 and f_2 are ϕ^{finite}-conjugate as elements of F. From this it follows that $\mathcal{R}(v)$ has cardinality $R(\phi^{finite})$. Since this is independent of v, we have

$$R(\phi) = \sum_v R(\phi^{finite}) = \mid \det(1-M) \mid \times R(\phi^{finite}).$$

Now consider the map $\phi^{\infty} : G/F \to G/F$. We have

$$\phi^{\infty}((v,F)) = (M \cdot v, \psi(v)F) = (M \cdot v, F).$$

From this it follows that ϕ^{∞} is isomorphic to map $M : \mathbf{Z}^k \to \mathbf{Z}^k$. This implies

$$R(\phi^{\infty}) = R(M : \mathbf{Z}^k \to \mathbf{Z}^k)$$

but it is known [31] that $R(M : \mathbf{Z}^k \to \mathbf{Z}^k) = \mid \det(1-M) \mid$. Therefore $R(\phi) = R(\phi^{finite}) \times R(\phi^{\infty})$, proving proposition 3 .

Let W be the complex vector space of complex valued class functions on the group F. The map ϕ induces a linear map $B : W \to W$ defined as above in theorem 15.

Theorem 18 *If G is the direct sum of a free Abelian and a finite group and ϕ an endomorphism of G . Then we have*

$$R(\phi) = (-1)^{r+p} \sum_{i=0}^{k} (-1)^i \mathrm{Tr}\, (\Lambda^i \phi^\infty \otimes B). \qquad (2.22)$$

where k is $rg(G/F)$, p the number of $\mu \in \mathrm{Spec}\, \phi^\infty$ such that $\mu < -1$, and r the number of real eigenvalues of ϕ^∞ whose absolute value is > 1.

PROOF Theorem follows from lemmas 15 and theorem 15 , proposition 3 and formula
$$\mathrm{Tr}\, (\Lambda^i \phi^\infty) \cdot \mathrm{Tr}\, (B) = \mathrm{Tr}\, (\Lambda^i \phi^\infty \otimes B).$$

Theorem 19 *Let G is the direct sum of free Abelian and a finite group and ϕ an endomorphism of G . If the numbers $R(\phi^n)$ are all finite then $R_\phi(z)$ is a rational function and is equal to*

$$R_\phi(z) = \left(\prod_{i=0}^{k} \det(1 - \Lambda^i \phi^\infty \otimes B \cdot \sigma \cdot z)^{(-1)^{i+1}} \right)^{(-1)^r} \qquad (2.23)$$

where matrix B is defined in theorem 15, $\sigma = (-1)^p$,p , r and k are constants described in lemma 17 .

PROOF From proposition 3 it follows that $R(\phi^n) = R((\phi^\infty)^n \cdot R((\phi^{finite})^n)$. From now on the proof repeat the proof of the theorem 12.

Corollary 6 *Let the assumptions of theorem 19 hold. Then the poles and zeros of the Reidemeister zeta function are complex numbers which are the reciprocal of an eigenvalues of one of the matrices*

$$\Lambda^i(\phi^\infty) \otimes B \cdot \sigma \qquad 0 \leq i \leq \mathrm{rank}\, G$$

Another proof of rationality of the Reidemeister zeta function gives

Theorem 20 *If G is the direct sum of a finitely generated free Abelian and a finite group and ϕ an endomorphism of G then $R_\phi(z)$ is a rational function and is equal to the following additive convolution:*

$$R_\phi(z) = R_\phi^\infty(z) * R_\phi^{finite}(z). \qquad (2.24)$$

where $R_\phi^\infty(z)$ is the Reidemeister zeta function of the endomorphism ϕ^∞ : $G^\infty \to G^\infty$, and $R_\phi^{finite}(z)$ is the Reidemeister zeta function of the endomorphism $\phi^{finite} : G^{finite} \to G^{finite}$. The functions $R_\phi^\infty(z)$ and $R_\phi^{finite}(z)$ are given by the formulae

$$R_\phi^\infty(z) = \left(\prod_{i=0}^{k} \det(1 - \Lambda^i \phi^\infty \cdot \sigma z)^{(-1)^{i+1}} \right)^{(-1)^r}, \qquad (2.25)$$

$$R_\phi^{finite}(z) = \prod_{[<g>]} \frac{1}{1 - z^{\#[<g>]}},$$

The product here is over all periodic ϕ-orbits of ordinary conjugacy classes of elements of G. The number $\#[< g >]$ is the number of conjugacy classes in the ϕ-orbit of the conjugacy class $< g >$. Also, $\sigma = (-1)^p$ where p is the number of real eingevalues $\lambda \in$ Spec ϕ^∞ such that $\lambda < -1$ and r is the number of real eingevalues $\lambda \in$ Spec ϕ^∞ such that $| \lambda |> 1$.

PROOF From proposition 3 it follows that $R(\phi^n) = R((\phi^\infty)^n) \cdot R((\phi^{finite})^n)$. From this we have

$$R_\phi(z) = R_{(\phi^\infty)}(z) * R_\phi^{finite}(z).$$

The rationality of $R_\phi(z)$ and the formulae for $R_\phi^\infty(z)$ and $R_\phi^{finite}(z)$ follow from the lemma 13 , lemma 17 and theorem 16 .

Theorem 21 (Functional equation) *Let $\phi : G \to G$ be an endomorphism of a group G which is the direct sum of a finitely generated free Abelian and a finite group. If G is finite the functional equation of R_ϕ is described in theorem 16. If G is infinite then R_ϕ has the following functional equation:*

$$R_\phi \left(\frac{1}{dz} \right) = \epsilon_2 \cdot R_\phi(z)^{(-1)^{\text{Rank } G}}. \qquad (2.26)$$

where $d = \det (\phi^\infty : G^\infty \to G^\infty)$ and ϵ_2 are a constants in $\mathbb{C}^\times$.

PROOF From proposition 3 we have $R_\phi(z) = R_\phi^\infty(z) * R_\phi^{finite}(z)$. In the lemma 19 and theorem 16 we have obtained functional equations for the functions $R_\phi^\infty(z)$ and $R_\phi^{finite}(z)$. Now, lemma 14 gives us the functional equation for $R_\phi(z)$.

2.6   Endomorphisms of nilpotent groups

In this section we consider finitely generated torsion free nilpotent group Γ.It is well known [63] that such group Γ is a uniform discrete subgroup of a simply connected nilpotent Lie group G (uniform means that the coset space G/Γ is compact).The coset space $M = G/\Gamma$ is called a nilmanifold.Since $\Gamma = \pi_1(M)$ and M is a $K(\Gamma, 1)$, every endomorphism $\phi : \Gamma \to \Gamma$ can be realized by a selfmap $f : M \to M$ such that $f_* = \phi$ and thus $R(f) = R(\phi)$.Any endomorphism $\phi : \Gamma \to \Gamma$ can be uniquely extended to an endomorphism $F : G \to G$.Let $\tilde{F} : \tilde{G} \to \tilde{G}$ be the corresponding Lie algebra endomorphism induced from F.

Theorem 22 *If Γ is a finitely generated torsion free nilpotent group and ϕ an endomorphism of Γ .Then*

$$R(\phi) = (-1)^{r+p} \sum_{i=0}^{m} (-1)^i \text{Tr } \Lambda^i \tilde{F}, \tag{2.27}$$

where m is $rg\Gamma = \dim M$, p the number of $\mu \in \text{Spec } \tilde{F}$ such that $\mu < -1$, and r the number of real eigenvalues of $\tilde{F}$ whose absolute value is > 1.

PROOF: Let $f : M \to M$ be a map realizing ϕ on a compact nilmanifold M of dimension m.We suppose that the Reidemeister number $R(f) = R(\phi)$ is finite.The finiteness of $R(f)$ implies the nonvanishing of the Lefschetz number $L(f)$ [36].A strengthened version of Anosov's theorem [4] is proven in [71] which states, in particular, that if $L(f) \neq 0$ than $N(f) = |L(f)| = R(f)$.It is well known that $L(f) = \det(\tilde{F} - 1)$ [4].From this we have

$$R(\phi) = R(f) = |L(f)| = |\det(1 - \tilde{F})| = (-1)^{r+p} \det(1 - \tilde{F}) =$$

$$= (-1)^{r+p} \sum_{i=0}^{m} (-1)^i \text{Tr } \Lambda^i \tilde{F}.$$

Theorem 23 *If Γ is a finitely generated torsion free nilpotent group and ϕ an endomorphism of Γ .Then $R_\phi(z)$ is a rational function and is equal to*

$$R_\phi(z) = \left(\prod_{i=0}^{m} \det(1 - \Lambda^i \tilde{F} \cdot \sigma \cdot z)^{(-1)^{i+1}} \right)^{(-1)^r} \tag{2.28}$$

where $\sigma = (-1)^p$,p , r, m and $\tilde{F}$ is defined in theorem 22.

PROOF If we repeat the proof of the previous theorem for ϕ^n instead ϕ we obtain that $R(\phi^n) = (-1)^{r+pn} \det(1 - \tilde{F})($ we suppose that Reidemeister numbers $R(\phi^n)$ are finite for all n).Last formula implies the trace formula for $R(\phi^n)$:

$$R(\phi^n) = (-1)^{r+pn} \sum_{i=0}^{m} (-1)^i \text{Tr } (\Lambda^i \tilde{F})^n$$

Now theorem follows immediately by direct calculation as in lemma 17.

Corollary 7 *Let the assumptions of theorem 23 hold. Then the poles and zeros of the Reidemeister zeta function are complex numbers which are reciprocal of an eigenvalue of one of the matrices*

$$\Lambda^i(\tilde{F}) : \Lambda^i(\tilde{G}) \longrightarrow \Lambda^i(\tilde{G}) \qquad 0 \leq i \leq \text{rank } G$$

2.6.1   Functional equation

Theorem 24 *Let $\phi : \Gamma \to \Gamma$ be an endomorphism of a finitely generated torsion free nilpotent group Γ. Then the Reidemeister zeta function $R_\phi(z)$ has the following functional equation:*

$$R_\phi \left(\frac{1}{dz} \right) = \epsilon_1 \cdot R_\phi(z)^{(-1)^{\text{Rank } G}}. \tag{2.29}$$

where $d = \det \tilde{F}$ and ϵ_1 are a constants in $\mathbb{C}^\times$.

PROOF Via the natural nonsingular pairing $(\Lambda^i \tilde{F}) \otimes (\Lambda^{m-i} \tilde{F}) \to \mathbb{C}$ the operators $\Lambda^{m-i} \tilde{F}$ and $d.(\Lambda^i \tilde{F})^{-1}$ are adjoint to each other.

We consider an eigenvalue λ of $\Lambda^i \tilde{F}$. By theorem 23, This contributes a term

$$\left((1 - \frac{\lambda\sigma}{dz})^{(-1)^{i+1}} \right)^{(-1)^r}$$

to $R_\phi \left(\frac{1}{dz} \right)$.

We rewrite this term as

$$\left(\left(1 - \frac{d\sigma z}{\lambda} \right)^{(-1)^{i+1}} \left(\frac{-dz}{\lambda\sigma} \right)^{(-1)^i} \right)^{(-1)^r}$$

and note that $\frac{d}{\lambda}$ is an eigenvalue of $\Lambda^{m-i}\tilde{F}$. Multiplying these terms together we obtain,

$$R_\phi\left(\frac{1}{dz}\right) = \left(\prod_{i=1}^{m} \prod_{\lambda^{(i)} \in \text{Spec } \Lambda^i \tilde{F}} \left(\frac{1}{\lambda^{(i)}\sigma}\right)^{(-1)^i}\right)^{(-1)^r} \times R_\phi(z)^{(-1)^m}.$$

The variable z has disappeared because

$$\sum_{i=0}^{m}(-1)^i \dim \Lambda^i \tilde{G} = \sum_{i=0}^{m}(-1)^i {C_k}^i = 0.$$

2.7 The Reidemeister zeta function and group extensions.

Suppose we are given a commutative diagram

$$
\begin{array}{ccc}
G & \xrightarrow{\phi} & G \\
\downarrow p & & \downarrow p \\
\overline{G} & \xrightarrow{\overline{\phi}} & \overline{G}
\end{array}
\tag{2.30}
$$

of groups and homomorphisms. In addition let the sequence

$$0 \to H \to G \xrightarrow{p} \overline{G} \to 0 \tag{2.31}$$

be exact. Then ϕ restricts to an endomorphism $\phi\mid_H\colon H \to H$.

Definition 11 *The short exact sequence (2.31) of groups is said to have a normal splitting if there is a section $\sigma : \overline{G} \to G$ of p such that $\text{Im } \sigma = \sigma(\overline{G})$ is a normal subgroup of G. An endomorphism $\phi : G \to G$ is said to preserve this normal splitting if ϕ induces a morphism of (2.31) with $\phi(\sigma(\overline{G})) \subset \sigma(\overline{G})$.*

In this section we study the relation between the Reidemeister zeta functions $R_\phi(z)$, $R_{\overline{\phi}}(z)$ and $R_{\phi\mid_H}(z)$.

Theorem 25 *Let the sequence (2.31) have a normal splitting which is preserved by $\phi : G \to G$. Then we have*

$$R_\phi(z) = R_{\overline{\phi}}(z) * R_{\phi\mid_H}(z).$$

In particular, if $R_{\overline{\phi}}(z)$ and $R_{\phi|_H}(z)$ are rational functions then so is $R_\phi(z)$. If $R_{\overline{\phi}}(z)$ and $R_{\phi|_H}(z)$ are rational functions with functional equations as described in theorems 21 and 24 then so is $R_\phi(z)$.

PROOF From the assumptions of the theorem it follows that for every $n > 0$

$$R(\phi^n) = R(\overline{\phi}^n) \cdot R(\phi^n |_H) \quad \text{(see [48]).}$$

2.8   The Reidemeister zeta function of a continuous map

Using Corrolary 1 from Chapter 1 we may apply all theorems about the Reidemeister zeta function of group endomorphisms to the Reidemeister zeta functions of continuous maps. Theorem 16 yield

Theorem 26 *Let X be a polyhedron with finite fundamental group $\pi_1(X)$ and let $f : X \to X$ be a continuous map. Then $R_f(z)$ is a rational function with a functional equation:*

$$R_f(z) = R_{\tilde{f}_*}(z) = \prod_{[<g>]} \frac{1}{1 - z^{\#[<g>]}},$$

$$R_f\left(\frac{1}{z}\right) = (-1)^a z^b R_f(z).$$

The product in the first formula is over all periodic $\tilde{f}_*$-orbits of ordinary conjugacy classes of elements of $\pi_1(X)$. The number $\#[< g >]$ is the number of conjugacy classes in the $\tilde{f}_*$-orbit of $< g >$. In the functional equation the numbers a and b are respectively the number of periodic $\tilde{f}_*$- orbits of cojugacy classes of elements of $\pi_1(X)$, and the number of periodic conjugacy classes of elements of $\pi_1(X)$.

Theorem 10 yield

Theorem 27 *Let $f : X \to X$ be eventually commutative. Then $R_f(z)$ is a rational function and is given by:*

$$R_f(z) = R_{\tilde{f}_*}(z) = R_{f_{1*}}(z) = R_{f_{1*}}^\infty(z) * R_{f_{1*}}^{finite}(z),$$

*where $R_{f_{1*}}^\infty(z)$ is the Reidemeister zeta function of the endomorphism f_{1*}^∞ :
$H_1(X, \mathbb{Z})^\infty \to H_1(X, \mathbb{Z})^\infty$ and $R_{f_{1*}}^{finite}(z)$ is the Reidemeister zeta function of
the endomorphism $f_{1*}^{finite} : H_1(X, \mathbb{Z})^{finite} \to H_1(X, \mathbb{Z})^{finite}$. The functions
$R_{f_{1*}}^\infty(z)$ and $R_{f_{1*}}^{finite}(z)$ are given by the formulae:*

$$R_{f_{1*}}^\infty(z) = \left(\prod_{i=0}^{k} \det \left(1 - \Lambda^i f_{1*}^\infty \sigma z \right)^{(-1)^{i+1}} \right)^{(-1)^r}$$

$$R_{f_{1*}}^{finite}(z) = \prod_{[h]} \frac{1}{1 - z^{\#[h]}}$$

*With the product over $[h]$ being taken over all periodic f_{1*}- orbits of torsion
elements $h \in H_1(X, \mathbb{Z})$, and with $\sigma = (-1)^p$ where p is the number of $\mu \in$
Spec f_{1*}^∞ such that $\mu < -1$. The number r is the number of real eigenvalues
of f_{1*}^∞ whose absolute value is > 1.*

Theorem 13 yield

Theorem 28 (Connection with Lefschetz zeta function) *Let $f : X \to
X$ be eventually commutative. Then*

$$R_f(z) = R_{\tilde{f}_*}(z) = R_{f_{1*}}(z) = L_{\widehat{(f_{1*})}}(\sigma z)^{(-1)^r},$$

*where r and σ are constants as described in theorem 27. If X is a polyhedron
with finite fundamental group then this reduces to*

$$R_f(z) = R_{\tilde{f}_*}(z) = R_{f_{1*}}(z) = L_{\widehat{(f_{1*})}}(z),$$

Theorem 29 (Functional equation) *Let $f : X \to X$ be eventually commutative. If $H_1(X; \mathbb{Z})$ is finite, then $R_f(z)$ has the following functional
equation:*

$$R_f \left(\frac{1}{z} \right) = (-1)^p z^q R_f(z),$$

*where p is the number of periodic orbits of f_{1*} in $H_1(X; \mathbb{Z})$ and q is the
number of periodic elements of f_{1*} in $H_1(X; \mathbb{Z})$.*

If $H_1(X; \mathbb{Z})$ is infinite then $R_f(z)$ has the following functional equation:

$$R_f \left(\frac{1}{dz} \right) = \epsilon_2 . R_f(z)^{(-1)^{\text{Rank } H_1(X;\mathbb{Z})}},$$

where $d = \det(f_{1*}^\infty) \in \mathbb{C}^\times$ and $\epsilon_2 \in \mathbb{C}^\times$ is a constant.

Theorem 20 yield

Theorem 30 *Let X be a polyhedron whose fundamental group π is the direct sum of a finitely generated free Abelian and a finite group. Let $f : X \to X$ be a continuous map. Then $R_f(z)$ is a rational function and is equal to the following additive convolution:*

$$R_f(z) = R_{\tilde{f}_*}^{\infty}(z) * R_{\tilde{f}_*}^{finite}(z). \tag{2.32}$$

where $R_{\tilde{f}_*}^{\infty}(z)$ is the Reidemeister zeta function of the endomorphism $\tilde{f}_*^{\infty}$: $\pi^{\infty} \to \pi^{\infty}$, and $R_{\tilde{f}_*}^{finite}(z)$ is the Reidemeister zeta function of the endomorphism $\tilde{f}_*^{finite} : \pi^{finite} \to \pi^{finite}$. The functions $R_{\tilde{f}_*}^{\infty}(z)$ and $R_{\tilde{f}_*}^{finite}(z)$ are given by the formulae

$$R_{\tilde{f}_*}^{\infty}(z) = \left(\prod_{i=0}^{k} \det(1 - \Lambda^i \tilde{f}_*^{\infty} \cdot \sigma z)^{(-1)^{i+1}} \right)^{(-1)^r}, \tag{2.33}$$

$$R_{\tilde{f}_*}^{finite}(z) = \prod_{[<g>]} \frac{1}{1 - z^{\#[<g>]}},$$

The product here is over all periodic $\tilde{f}_*$-orbits of ordinary conjugacy classes of elements of π. The number $\#[< g >]$ is the number of conjugacy classes in the $\tilde{f}_*$-orbit of the conjugacy class $< g >$. Also, $\sigma = (-1)^p$ where p is the number of real eingevalues $\lambda \in \text{Spec } \tilde{f}_*^{\infty}$ such that $\lambda < -1$ and r is the number of real eingevalues $\lambda \in \text{Spec } \tilde{f}_*^{\infty}$ such that $| \lambda |> 1$.

Theorem 31 *Let $f : X \to X$ be a self map of of a nilmanifold. Then*

$$R_f(z) = \left(\prod_{i=0}^{m} \det \left(1 - \Lambda^i \tilde{F} \cdot \sigma \cdot z \right)^{(-1)^{i+1}} \right)^{(-1)^r}$$

$$R_f \left(\frac{1}{dz} \right) = \epsilon_1 \cdot R_f(z)^{(-1)^{\text{Rank } \pi_1(X)}},$$

where $\sigma = (-1)^p$,p , r, m and $\tilde{F}$ is defined in theorem 22, $d = \det \tilde{F}$ and ϵ_1 are a constants in $\mathbb{C}^{\times}$.

2.8.1 The Reidemeister zeta function of a continuous map and Serre bundles.

Let $p : E \to B$ be a Serre bundle in which E, B and every fibre are connected, compact polyhedra and $F_b = p^{-1}(b)$ is a fibre over $b \in B$. A Serre bundle $p : E \to B$ is said to be *(homotopically) orientable* if for any two paths w, w' in B with the same endpoints $w(0) = w'(0)$ and $w(1) = w'(1)$, the fibre translations $\tau_w, \tau_{w'} : F_{w(0)} \to F_{w(1)}$ are homotopic. A map $f : E \to E$ is called a *fibre map* if there is an induced map $\bar{f} : B \to B$ such that $p \circ f = \bar{f} \circ p$. Let $p : E \to B$ be an orientable Serre bundle and let $f : E \to E$ be a fibre map. Then for any two fixed points b, b' of $\bar{f} : B \to B$ the maps $f_b = f \mid_{F_b}$ and $f_{b'} = f \mid_{F_{b'}}$ have the same homotopy type; hence they have the same Reidemeister numbers $R(f_b) = R(f_{b'})$ [51].

The following theorem describes the relation between the Reidemeister zeta functions $R_f(z)$, $R_{\bar{f}}(z)$ and $R_{f_b}(z)$ for a fibre map $f : E \to E$ of an orientable Serre bundle $p : E \to B$.

Theorem 32 *Suppose that $f : E \to E$ admits a Fadell splitting in the sense that for some e in* Fix f *and $b = p(e)$ the following conditions are satisfied:*

1. *the sequence*

$$0 \longrightarrow \pi_1(F_b, e) \xrightarrow{i_*} \pi_1(E, e) \xrightarrow{p_*} \pi_1(B, e) \longrightarrow 0$$

   is exact,

2. *p_* admits a right inverse (section) σ such that* Im σ *is a normal subgroup of $\pi_1(E, e)$ and $f_*(\text{Im } \sigma) \subset \text{Im } \sigma$.*

We then have

$$R_f(z) = R_{\bar{f}}(z) * R_{f_b}(z).$$

If $R_{\bar{f}}(z)$ and $R_{f_b}(z)$ are rational functions then so is $R_f(z)$. If $R_{\bar{f}}(z)$ and $R_{f_b}(z)$ are rational functions with functional equations as described in theorems 26 and 29 then so is $R_f(z)$.

PROOF The proof follows from theorem 25 .

Chapter 3

The Nielsen zeta function

3.1   Radius of Convergence of the Nielsen zeta function

In this section we find a sharp estimate for the radius of convergence of the Nielsen zeta function in terms of the topological entropy of the map. It follows from this estimate that the Nielsen zeta function has positive radius of convergence.

3.1.1   Topological entropy

The most widely used measure for the complexity of a dynamical system is the topological entropy. For the convenience of the reader, we include its definition. Let $f : X \to X$ be a self-map of a compact metric space. For given $\epsilon > 0$ and $n \in \mathbb{N}$, a subset $E \subset X$ is said to be (n, ϵ)-separated under f if for each pair $x \neq y$ in E there is $0 \leq i < n$ such that $d(f^i(x), f^i(y)) > \epsilon$. Let $s_n(\epsilon, f)$ denote the largest cardinality of any (n, ϵ)-separated subset E under f. Thus $s_n(\epsilon, f)$ is the greatest number of orbit segments $x, f(x), ..., f^{n-1}(x)$ of length n that can be distinguished one from another provided we can only distinguish between points of X that are at least ϵ apart. Now let

$$h(f, \epsilon) := \limsup_n \frac{1}{n} \cdot \log \, s_n(\epsilon, f)$$

$$h(f) := \limsup_{\epsilon \to 0} h(f, \epsilon).$$

60

The number $0 \leq h(f) \leq \infty$, which to be independent of the metric d used, is called the topological entropy of f. If $h(f, \epsilon) > 0$ then, up to resolution $\epsilon > 0$, the number $s_n(\epsilon, f)$ of distinguishable orbit segments of length n grows exponentially with n. So $h(f)$ measures the growth rate in n of the number of orbit segments of length n with arbitrarily fine resolution. A basic relation between Nielsen numbers and topological entropy was found by N.Ivanov [55] and independently by Aronson and Grines. We present here a very short proof of Jiang [52] of the Ivanov's inequality.

Lemma 22 *[55]*

$$h(f) \geq \limsup_n \frac{1}{n} \cdot \log N(f^n)$$

PROOF Let δ be such that every loop in X of diameter $< 2\delta$ is contractible. Let $\epsilon > 0$ be a smaller number such that $d(f(x), f(y)) < \delta$ whenever $d(x, y) < 2\epsilon$. Let $E_n \subset X$ be a set consisting of one point from each essential fixed point class of f^n. Thus $\mid E_n \mid = N(f^n)$. By the definition of $h(f)$, it suffices to show that E_n is (n, ϵ)-separated. Suppose it is not so. Then there would be two points $x \neq y \in E_n$ such that $d(f^i(x), f^i(y)) \leq \epsilon$ for $o \leq i < n$ hence for all $i \geq 0$. Pick a path c_i from $f^i(x)$ to $f^i(y)$ of diameter $< 2\epsilon$ for $o \leq i < n$ and let $c_n = c_0$. By the choice of δ and ϵ , $f \circ c_i \simeq c_{i+1}$ for all i, so $f^n \circ c_0 \simeq c_n = c_0$. This means x, y in the same fixed point class of f^n, contradicting the construction of E_n.

This inequality is remarkable in that it does not require smoothness of the map and provides a common lower bound for the topological entropy of all maps in a homotopy class.

We denote by R the radius of convergence of the Nielsen zeta function $N_f(z)$. Let $h = \inf h(g)$ over all maps g of the same homotopy type as f.

Theorem 33 *For a continuous map of a compact polyhedron X into itself,*

$$R \geq \exp(-h) > 0. \tag{3.1}$$

PROOF The inequality $R \geq \exp(-h)$ follows from the previous lemma, the Cauchy-Hadamard formula, and the homotopy invariance of the radius R of the Nielsen zeta function $N_f(z)$. We consider a smooth compact manifold M which is a regular neighborhood of X and a smooth map $g : M \to M$ of the same homotopy type as f. As is known [75] , the entropy $h(g)$ is finite. Thus $\exp(-h) \geq \exp(-h(g)) > 0$.

3.1.2 Algebraic lower estimation for the Radius of Convergence

In this subsection we propose another prove of positivity of the radius R and give an exact algebraic lower estimation for the radius R using the Reidemeister trace formula for generalized Lefschetz numbers.

The fundamental group $\pi = \pi_1(X, x_0)$ splits into $\tilde{f}_*$-conjugacy classes.Let π_f denote the set of $\tilde{f}_*$-conjugacy classes,and $\mathbb{Z}\pi_f$ denote the abelian group freely generated by π_f . We will use the bracket notation $a \to [a]$ for both projections $\pi \to \pi_f$ and $\mathbb{Z}\pi \to \mathbb{Z}\pi_f$. Let x be a fixed point of f.Take a path c from x_0 to x.The $\tilde{f}_*$-conjugacy class in π of the loop $c \cdot (f \circ c)^{-1}$,which is evidently independent of the choice of c, is called the coordinate of x.Two fixed points are in the same fixed point class F iff they have the same coordinates.This $\tilde{f}_*$-conjugacy class is thus called the coordinate of the fixed point class F and denoted $cd_\pi(F, f)$ (compare with description in section 1).

The generalized Lefschetz number or the Reidemeister trace [78] is defined as

$$L_\pi(f) := \sum_F \text{Index } (F, f).cd_\pi(F, f) \in \mathbb{Z}\pi_f, \qquad (3.2)$$

the summation being over all essential fixed point classes F of f.The Nielsen number $N(f)$ is the number of non-zero terms in $L_\pi(f)$,and the indices of the essential fixed point classes appear as the coefficients in $L_\pi(f)$.This invariant used to be called the Reidemeister trace because it can be computed as an alternating sum of traces on the chain level as follows [78],[97] . Assume that X is a finite cell complex and $f : X \to X$ is a cellular map.A cellular decomposition e_j^d of X lifts to a π-invariant cellular structure on the universal covering $\tilde{X}$.Choose an arbitrary lift $\tilde{e}_j^d$ for each e_j^d . They constitute a free $\mathbb{Z}\pi$-basis for the cellular chain complex of $\tilde{X}$.The lift $\tilde{f}$ of f is also a cellular map.In every dimension d, the cellular chain map $\tilde{f}$ gives rise to a $\mathbb{Z}\pi$-matrix $\tilde{F}_d$ with respect to the above basis,i.e $\tilde{F}_d = (a_{ij})$ if $\tilde{f}(\tilde{e}_i^d) = \sum_j a_{ij}\tilde{e}_j^d$,where $a_{ij} \in \mathbb{Z}\pi$.Then we have the Reidemeister trace formula

$$L_\pi(f) = \sum_d (-1)^d [\text{Tr } \tilde{F}_d] \in \mathbb{Z}\pi_f. \qquad (3.3)$$

Now we describe alternative approach to the Reidemeister trace formula proposed recently by Jiang [52]. This approach is useful when we study the periodic points of f, i.e. the fixed points of the iterates of f.

The mapping torus T_f of $f : X \to X$ is the space obtained from $X \times [o, \infty)$ by identifying $(x, s + 1)$ with $(f(x), s)$ for all $x \in X, s \in [0, \infty)$. On T_f there is a natural semi-flow $\phi : T_f \times [0, \infty) \to T_f, \phi_t(x, s) = (x, s + t)$ for all $t \geq 0$. Then the map $f : X \to X$ is the return map of the semi-flow ϕ. A point $x \in X$ and a positive number $\tau > 0$ determine the orbit curve $\phi_{(x, \tau)} := \phi_t(x)_{0 \leq t \leq \tau}$ in T_f.

Take the base point x_0 of X as the base point of T_f. It is known that the fundamental group $H := \pi_1(T_f, x_0)$ is obtained from π by adding a new generator z and adding the relations $z^{-1}gz = \tilde{f}_*(g)$ for all $g \in \pi = \pi_1(X, x_0)$. Let H_c denote the set of conjugacy classes in H. Let $\mathbb{Z}H$ be the integral group ring of H, and let $\mathbb{Z}H_c$ be the free abelian group with basis H_c. We again use the bracket notation $a \to [a]$ for both projections $H \to H_c$ and $\mathbb{Z}H \to \mathbb{Z}H_c$. If F^n is a fixed point class of f^n, then $f(F^n)$ is also fixed point class of f^n and Index $(f(F^n), f^n) =$ Index (F^n, f^n). Thus f acts as an index-preserving permutation among fixed point classes of f^n. By definition, an n-orbit class O^n of f to be the union of elements of an orbit of this action. In other words, two points $x, x' \in$ Fix (f^n) are said to be in the same n-orbit class of f if and only if some $f^i(x)$ and some $f^j(x')$ are in the same fixed point class of f^n. The set Fix (f^n) splits into a disjoint union of n-orbits classes. Point x is a fixed point of f^n or a periodic point of period n if and only if orbit curve $\phi_{(x,n)}$ is a closed curve. The free homotopy class of the closed curve $\phi_{(x,n)}$ will be called the H-coordinate of point x, written $cd_H(x, n) = [\phi_{(x,n)}] \in H_c$. It follows that periodic points x of period n and x' of period n' have the same H-coordinate if and only if $n = n'$ and x, x' belong to the same n-orbits class of f. Thus it is possible equivalently define $x, x' \in$ Fix f^n to be in the same n-orbit class if and only if they have the same H-coordinate.

Recently, Jiang [52] has considered generalized Lefschetz number with respect to H

$$L_H(f^n) := \sum_{O^n} \text{Index } (O^n, f^n) \cdot cd_H(O^n) \in \mathbb{Z}H_c, \qquad (3.4)$$

and proved following trace formula:

$$L_H(f^n) = \sum_d (-1)^d [\text{Tr } (z\tilde{F}_d)^n] \in \mathbb{Z}H_c, \qquad (3.5)$$

where $\tilde{F}_d$ be $\mathbb{Z}\pi$-matrices defined above and $z\tilde{F}_d$ is regarded as a $\mathbb{Z}H$-matrix.

For any set S let $\mathbb{Z}S$ denote the free abelian group with the specified basis S. The norm in $\mathbb{Z}S$ is defined by

$$\|\sum_i k_i s_i\| := \sum_i \mid k_i \mid \in \mathbb{Z}, \tag{3.6}$$

when the s_i in S are all different.

For a $\mathbb{Z}H$-matrix $A = (a_{ij})$, define its norm by $\|A\| := \sum_{i,j}\|a_{ij}\|$. Then we have inequalities $\|AB\| \le \|A\| \cdot \|B\|$ when A, B can be multiplied, and $\|\mathrm{Tr}\ A\| \le \|A\|$ when A is a square matrix. For a matrix $A = (a_{ij})$ in $\mathbb{Z}S$, its matrix of norms is defined to be the matrix $A^{norm} := (\|a_{ij}\|)$ which is a matrix of non-negative integers. In what follows, the set S will be π, H or H_c. We denote by $s(A)$ the spectral radius of A, $s(A) = \lim_n(\|A^n\|)^{\frac{1}{n}}$, which coincide with the largest module of an eigenvalue of A.

Theorem 34 *For any continuous map f of any compact polyhedron X into itself the Nielsen zeta function has positive radius of convergence R, which admits following estimations*

$$R \ge \frac{1}{\max_d \|z\tilde{F}_d\|} > 0 \tag{3.7}$$

and

$$R \ge \frac{1}{\max_d s(\tilde{F}_d^{norm})} > 0, \tag{3.8}$$

PROOF By the homotopy type invariance of the invariants we can suppose that f is a cell map of a finite cell complex. By the definition, the Nielsen number $N(f^n)$ is the number of non-zero terms in $L_\pi(f^n)$. The norm $\|L_H(f^n)\|$ is the sum of absolute values of the indices of all the n-orbits classes O^n. It equals $\|L_\pi(f^n)\|$, the sum of absolute values of the indices of all the fixed point classes of f^n, because any two fixed point classes of f^n contained in the same n-orbit class O^n must have the same index. From this we have

$$N(f^n) \le \|L_\pi(f^n)\| = \|L_H(f^n)\| = \|\sum_d(-1)^d[\mathrm{Tr}\ (z\tilde{F}_d)^n]\| \le$$

$$\le \sum_d \|[\mathrm{Tr}\ (z\tilde{F}_d)^n]\| \le \sum_d \|\mathrm{Tr}\ (z\tilde{F}_d)^n\| \le \sum_d \|(z\tilde{F}_d)^n\| \le \sum_d \|(z\tilde{F}_d)\|^n$$

(see [52]).The radius of convergence R is given by Caushy-Adamar formula:

$$\frac{1}{R} = \limsup_{n} \left(\frac{N(f^n)}{n}\right)^{\frac{1}{n}} = \limsup_{n} (N(f^n))^{\frac{1}{n}}.$$

Therefore we have:

$$R = \frac{1}{\limsup_{n}(N(f^n))^{\frac{1}{n}}} \geq \frac{1}{\max_d \|z\tilde{F}_d\|} > 0.$$

Inequalities:

$$N(f^n) \leq \|L_\pi(f^n)\| = \|L_H(f^n)\| = \|\sum_d (-1)^d [\text{Tr} \ (z\tilde{F}_d)^n]\| \leq \sum_d \|[\text{Tr} \ (z\tilde{F}_d)^n]\| \leq$$

$$\leq \sum_d \|\text{Tr} \ (z\tilde{F}_d)^n\| \leq \sum_d \text{Tr} \ ((z\tilde{F}_d)^n)^{norm} \leq \sum_d \text{Tr} \ ((z\tilde{F}_d)^{norm})^n \leq$$

$$\leq \sum_d \text{Tr} \ ((\tilde{F}_d)^{norm})^n$$

and the definition of spectral radius give estimation:

$$R = \frac{1}{\limsup_{n}(N(f^n))^{\frac{1}{n}}} \geq \frac{1}{\max_d s(\tilde{F}_d^{norm})} > 0.$$

Example 7 *Let X be surface with boundary, and $f : X \to X$ be a map.Fadell and Husseini [20] devised a method of computing the matrices of the lifted chain map for surface maps.Suppose $\{a_1, ..., a_r\}$ is a free basis for $\pi_1(X)$. Then X has the homotopy type of a bouquet B of r circles which can be decomposed into one 0-cell and r 1-cells corresponding to the a_i,and f has the homotopy type of a cellular map $g : B \to B$. By the homotopy type invariance of the invariants,we can replace f with g in computations.The homomorphism $\tilde{f}_* : \pi_1(X) \to \pi_1(X)$ induced by f and g is determined by the images $b_i = \tilde{f}_*(a_i), i = 1, .., r$.The fundamental group $\pi_1(T_f)$ has a presentation $\pi_1(T_f) = < a_1, ..., a_r, z | a_i z = z b_i, i = 1, .., r >$.Let*

$$D = \left(\frac{\partial b_i}{\partial a_j}\right)$$

be the Jacobian in Fox calculus(see [10]).Then,as pointed out in [20], the matrices of the lifted chain map $\tilde{g}$ are

$$\tilde{F}_0 = (1), \tilde{F}_1 = D = \left(\frac{\partial b_i}{\partial a_j}\right).$$

Now, we can find estimations for the radius R from (3.7) and (3.8).

3.2   Nielsen zeta function of a periodic map

The following problem is of interest: for which spaces and classes of maps is the Nielsen zeta function rational? When is it algebraic? Can $N_f(z)$ be transcendental? Sometimes one can answer these questions without directly calculating the Nielsen numbers $N(f^n)$, but using the connection between Nielsen numbers of iterates. We denote $N(f^n)$ by N_n.We shall say that $f : X \to X$ is a periodic map of period m, if f^m is the identity map $id_X : X \to X$. Let $\mu(d), d \in N$, be the Möbius function of number theory. As is known, it is given by the following equations: $\mu(d) = 0$ if d is divisible by a square different from one ; $\mu(d) = (-1)^k$ if d is not divisible by a square different from one , where k denotes the number of prime divisors of d; $\mu(1) = 1$.

Theorem 35 *Let f be a periodic map of least period m of the connected compact polyhedron X . Then the Nielsen zeta function is equal to*

$$N_f(z) = \prod_{d|m} \sqrt[d]{(1 - z^d)^{-P(d)}},$$

where the product is taken over all divisors d of the period m, and $P(d)$ is the integer

$$P(d) = \sum_{d_1|d} \mu(d_1) N_{d|d_1}.$$

PROOF Since $f^m = id$, for each $j, N_j = N_{m+j}$. Since $(k, m) = 1$, there exist positive integers t and q such that $kt = mq + 1$. So $(f^k)^t = f^{kt} = f^{mq+1} = f^{mq}f = (f^m)^q f = f$. Consequently, $N((f^k)^t) = N(f)$. Let two fixed point x_0 and x_1 belong to the same fixed point class. Then there exists a path α from x_0 to x_1 such that $\alpha * (f \circ \alpha)^{-1} \simeq 0$. We have $f(\alpha * f \circ \alpha)^{-1}) = (f \circ \alpha) * (f^2 \circ \alpha)^{-1} \simeq 0$ and a product $\alpha * (f \circ \alpha)^{-1} * (f \circ \alpha) * (f^2 \circ \alpha)^{-1} = \alpha * (f^2 \circ \alpha)^{-1} \simeq 0$. It follows that $\alpha * (f^k \circ \alpha)^{-1} \simeq 0$ is derived by the iteration of this process.So x_0 and x_1 belong to the same fixed point class of f^k. If two point belong to the different fixed point classes f, then they belong to the different fixed point classes of f^k . So, each essential class(class with nonzero index) for f is an essential class for f^k; in addition , different essential classes for f are different essential classes for f^k. So $N(f^k) \geq N(f)$. Analogously, $N(f) = N((f^k)^t) \geq N(f^k)$. Consequently , $N(f) = N(f^k)$. One can prove completely analogously that $N_d = N_{di}$, if $(i, m/d) = 1$, where

d is a divisor of m. Using these series of equal Nielsen numbers, one can regroup the terms of the series in the exponential of the Nielsen zeta function so as to get logarithmic functions by adding and subtracting missing terms with necessary coefficient. We show how to do this first for period $m = p^l$, where p is a prime number . We have the following series of equal Nielsen numbers:

$$N_1 = N_k, (k, p^l) = 1 (i.e., no\ N_{ip}, N_{ip^2}, ..., N_{ip^l}, i = 1, 2, 3, ...),$$

$$N_p = N_{2p} = N_{3p} = ... = N_{(p-1)p} = N_{(p+1)p} = ... (no\ N_{ip^2}, N_{ip^3}, ..., N_{ip^l})$$

etc.; finally,

$$N_{p^{l-1}} = N_{2p^{l-1}} = ... (no\ N_{ip^l})$$

and separately the number N_{p^l}.
Further,

$$\sum_{i=1}^{\infty} \frac{N_i}{i} z^i = \sum_{i=1}^{\infty} \frac{N_1}{i} z^i + \sum_{i=1}^{\infty} \frac{(N_p - N_1)}{p} \frac{z^{pi}}{i} +$$

$$+ \sum_{i=1}^{\infty} \frac{(N_{p^2} - (N_p - N_1) - N_1)}{p^2} \frac{z^{p^2 i}}{i} + ...$$

$$+ \sum_{i=1}^{\infty} \frac{(N_{p^l} - ... - (N_p - N_1) - N_1)}{p^l} \frac{z^{p^l i}}{i}$$

$$= -N_1 \cdot \log(1 - z) + \frac{N_1 - N_p}{p} \cdot \log(1 - z^p) +$$

$$+ \frac{N_p - N_{p^2}}{p^2} \cdot \log(1 - z^{p^2}) + ...$$

$$+ \frac{N_{p^{l-1}} - N_{p^l}}{p^l} \cdot \log(1 - z^{p^l}).$$

For an arbitrary period m , we get completely analogously,

$$N_f(z) = \exp\left(\sum_{i=1}^{\infty} \frac{N(f^i)}{i} z^i\right)$$

$$= \exp\left(\sum_{d|m} \sum_{i=1}^{\infty} \frac{P(d)}{d} \cdot \frac{z^{di}}{i}\right)$$

$$= \exp\left(\sum_{d|m} \frac{P(d)}{d} \cdot \log(1 - z^d)\right)$$

$$= \prod_{d|m} \sqrt[d]{(1 - z^d)^{-P(d)}}$$

where the integers $P(d)$ are calculated recursively by the formula

$$P(d) = N_d - \sum_{d_1|d; d_1 \neq d} P(d_1).$$

Moreover, if the last formula is rewritten in the form

$$N_d = \sum_{d_1|d} \mu(d_1) \cdot P(d_1)$$

and one uses the Möbius Inversion law for real function in number theory, then

$$P(d) = \sum_{d_1|d} \mu(d_1) \cdot N_{d/d_1},$$

where $\mu(d_1)$ is the Möbius function in number theory. The theorem is proved.

Corollary 8 *If in Theorem 35 the period m is a prime number, then*

$$N_f(z) = \frac{1}{(1 - z)^{N_1}} \cdot \sqrt[m]{(1 - z^m)^{N_1 - N_m}}.$$

For an involution of a connected compact polyhedron, we get

$$N_{in}(z) = \frac{1}{(1 - z)^{N_1}} \cdot \sqrt[2]{(1 - z^2)^{N_1 - N_2}}.$$

Remark 3 *Let $f : M^n \to M^n$, $n \geq 3$ be a homeomorphism of a compact hyperbolic manifold. Then by Mostow rigidity theorem f is homotopic to periodic homeomorphism g. So theorem 35 applies and the Nielsen zeta function $N_f(z)$ is equal to*

$$N_f(z) = N_g(z) = \prod_{d|m} \sqrt[d]{(1 - z^d)^{-P(d)}},$$

where the product is taken over all divisors d of the least period m of g, and $P(d)$ is the integer $P(d) = \sum_{d_1|d} \mu(d_1) N(g^{d|d_1})$.

Remark 4 *Let* $f : X \to X$ *be a continuous map of a connected compact poly-hedron* X, *homotopic to* id_X . *Since the Lefschetz numbers* $L(f^n) = L(id_X) = \chi(X)$, *where* $\chi(X)$ *is the Euler characteristic of* X, *then for* $\chi(X) \neq 0$ *one has* $N(f^n) = N(id_X) = 1$ *for all* $n > 0$, *and* $N_f(z) = \frac{1}{1-z}$; *if* $\chi(X) = 0$, *then* $N(f^n) = N(id_X) = 0$ *for all* $n > 0$, *and* $N_f(z) = 1$

3.3 Orientation-preserving homeomorphisms of a compact surface

The proof of the following theorem is based on Thurston's theory of orienta-tion-preserving homeomorphisms of surfaces [91].

Theorem 36 *The Nielsen zeta function of an orientation-preserving home-omorphism* f *of a compact surface* M^2 *is either a rational function or the radical of a rational function.*

PROOF The case of an orientable surface with $\chi(M^2) \geq 0(S^2, T^2)$ is con-sidered in subsection 3.8. In the case of an orientable surface with $\chi(M^2) < 0$, according to Thurston's classification theorem, the homeomorphism f is isotopic either to a periodic or a pseudo-Anosov, or a reducible homeomor-phism.In the first case the assertion of the theorem follows from theorem 35. If f is an orientation-preserving pseudo-Anosov homeomorphism of a compact surface(i.e. there is a number $\lambda > 1$ and a pair of transverse mea-sured foliations (F^s, μ^s) and (F^u, μ^u) such that $f(F^s, \mu^s) = (F^s, \frac{1}{\lambda}\mu^s)$ and $f(F^u, \mu^u) = (F^u, \lambda\mu^u))$, then for each $n > 0, N(f^n) = F(f^n)$ [91], [55], [51]. Consequently, in this case the Nielsen zeta function coincides with the Artin-Mazur zeta function: $N_f(z) = F_f(z)$. Since in [22] Markov partitions are constructed for a pseudo-Anosov homeomorphism, Manning's proof [64] of the rationality of the Artin-Mazur zeta function for diffeomorphisms satisfy-ing Smale's axiom A carries over to the case of pseudo-Anosov homeomor-phisms.Thus , the Nielsen zeta function $N_f(z)$ is also rational.Now if f is isotopic to a reduced homeomorphism ϕ, then there exists a reducing system S of disjoint circles $S_1, S_2, ..., S_m$ on $int M^2$ such that

1) each circle S_i does not bound a disk in M^2;

2) S_i is not isotopic to $S_j, i \neq j$;

3) the system of circles S is invariant with respect to ϕ;

4) the system S has an open ϕ-invariant tubular neighborhood $\eta(S)$ such that each ϕ -component Γ_j of the set $M^2 - \eta(S)$ is mapped into itself by some iterate $\phi^{n_j}, n_j > 0$ of the map ϕ; here ϕ^{n_j} on Γ_j is either a pseudo-Anosov or a periodic homeomorphism;

5) each band $\eta(S_i)$ is mapped into itself by some iterate $\phi^{m_i}, m_i > 0$; here ϕ^{m_i} on $\eta(S_i)$ is a generalized twist (possibly trivial).

Since the band $\eta(S_i)$ is homotopically equivalent to the circle S^1, as will be shown in subsection 3.8 the Nielsen zeta function $N_{\phi_i^m}(z)$ is rational. The zeta functions $N_\phi(z)$ and $N_{\phi^m}(z)$ are connected on the ϕ - component Γ_j by the formula $N_\phi(z) = \sqrt[n_j]{N_{\phi_j^n}(z^{n_j})}$; analogously, on the band $\eta(S_i), N_\phi(z) = \sqrt[m_j]{N_{\phi_j^m}(z^{m_j})}$. The fixed points of ϕ^n, belonging to different components Γ_j and bands $\eta(S_i)$ are nonequivalent [56],so the Nielsen number $N(\phi^n)$ is equal to the sum of the Nielsen numbers $N(\phi^n/\Gamma_j)$ and $N(\phi^n/\eta(S_i))$ of ϕ -components and bands . Consequently, by the properties of the exponential, the Nielsen zeta function $N_\phi(z) = N_f(z)$ is equal to the product of the Nielsen zeta functions of the ϕ- components Γ_j and the bands $\eta(S_i)$, i.e. is the radical of a rational function.

Remark 5 *For an orientation -preserving pseudo-Anosov homeomorphism of a compact surface the radius of the Nielsen zeta function $N_f(z)$ is equal to*

$$R = \exp(-h(f)) = \frac{1}{\lambda(A)},$$

where $\lambda(A)$ is the largest eigenvalue of the transition matrix of the topological Markov chain corresponding to f.

3.3.1 Geometry of the Mapping Torus and Radius of Convergence

Let $f : M^2 \to M^2$ be an orientation-preserving homeomorphism of a compact orientable surface, R the radius of convergence of Nielsen zeta function and T_f

be the mapping torus of f, i.e. T_f is obtained from $M^2 \times [0, 1]$ by identifying $(x, 0)$ to $(f(x), 1)$, $x \in M^2$.

Lemma 23 *Let the Euler characteristic $\chi(M^2) < 0$. Then $R = 1$ if and only if the Thurston normal form for f does not contain pseudo-Anosov components.*

PROOF Let ϕ be a Thurston canonical form of f. Suppose that ϕ does not contain a pseudo-Anosov component. Then ϕ is periodic or reducible by S, where on each ϕ-component ϕ is periodic. If ϕ is periodic then by theorem 35 $R = 1$. If ϕ is reducible then in theorem 36 we have proved that the Nielsen zeta function $N_\phi(z) = N_f(z)$ is equal to the product of the Nielsen zeta functions of the ϕ- components Γ_j and the bands $\eta(S_i)$, i.e. has radius of convergence $R = 1$.

A three-dimensional manifold M is called a graph manifold if there is a system of mutually disjoint two-dimensional tori $\{T_i\}$ in M such that the closure of each component of M cut along $\cup T_i$ is a (surface) $\times S^1$.

Theorem 37 *Let $\chi(M^2) < 0$. The mapping torus T_f is a graph-manifold if and only if $R = 1$. If $Int T_f$ admits a hyperbolic structure of finite volume , then $R < 1$. If $R < 1$ then f has an infinite set of periodic points with pairwise different periods.*

PROOF T. Kobayashi [58] has proved that mapping torus T_f is a graph-manifold if and only if the Thurston normal form for f does not contain pseudo-Anosov components.So, lemma 23 implies the first statement of the theorem. Thurston has proved [93], [90] that $Int T_f$ admits a hyperbolic structure of finite volume if and only if f is isotopic to pseudo-Anosov homeomorphism. But for pseudo-Anosov homeomorphism $1 > R = \exp(-h(f)) = \frac{1}{\lambda(A)}$, where $\lambda(A)$ is the largest eigenvalue of the transition matrix of the topological Markov chain corresponding to f.This proves the second statement of the theorem. If $R < 1$ thenThurston normal form for f does contain pseudo-Anosov components. It is known [58] that pseudo-Anosov homeomorphism has infinitely many periodic points those periods are mutually distinct.

A link L is a finite union of mutually disjoint circles in a three-dimensional manifold.The exterior of L is the closure of the complement of a regular neighborhood of L. A link L is a graph link if the exterior of L is a graph manifold.Let Σ a set consisting of a finite number of periodic orbits of f. The set $\Sigma \times [0, 1]$ projects to a link $L_{f,\Sigma}$ in the mapping torus T_f.

Corollary 9 *Let $\chi(M^2 - \Sigma < 0$. The link $L_{f,\Sigma}$ is a graph link if and only if $R = 1$.*

PROOF Homeomorphism f is isotope *relΣ* to a diffeomorphism g. Let F be a surface obtained from $M^2 - \Sigma$ by adding a circle to each end . Since g is differentiable at each point of Σ , g extends to $\tilde{g} : F \to F$ [47]. By theorem 37 $T_{\tilde{g}}$ is a graph manifold if and only if $R = 1$. Hence $L_{f,\Sigma}$ is a graph link if and only if $R = 1$.

3.4 The Jiang subgroup and the Nielsen zeta function

From the homotopy invariance theorem (see [51]) it follows that if a homotopy $\{h_t\} : f \cong g : X \to X$ lifts to a homotopy $\{\tilde{h}_t\} : \tilde{f} \cong \tilde{g} : \tilde{X} \to \tilde{X}$, then we have Index $(f, p(\text{Fix } \tilde{f})) = $ Index $(g, p(\text{Fix } \tilde{g}))$. Suppose $\{h_t\}$ is a cyclic homotopy $\{h_t\} : f \cong f$; then this lifts to a homotopy from a given lifting $\tilde{f}$ to another lifting $\tilde{f}' = \alpha \circ \tilde{f}$, and we have

$$\text{Index } (f, p(\text{Fix } \tilde{f})) = \text{Index } (f, p(\text{Fix } \alpha \circ \tilde{f})).$$

In other words, a cyclic homotopy induces a permutation of lifting classes (and hence of fixed point classes); those in the same orbit of this permutation have the same index. This idea is applied to the computation of $N_f(z)$.

Definition 12 *The trace subgroup of cyclic homotopies (the Jiang subgroup) $I(\tilde{f}) \subset \pi$ is defined by*

$$I(\tilde{f}) = \left\{ \alpha \in \pi \ \middle| \ \begin{array}{l} \text{there exists a cyclic homotopy} \\ \{h_t\} : f \cong f \text{which lifts to} \\ \{\tilde{h}_t\} : \tilde{f} \cong \alpha \circ \tilde{f} \end{array} \right\}$$

(see [51]).

Let $Z(G)$ denote the centre of a group G, and let $Z(H,G)$ denote the centralizer of the subgroup $H \subset G$. The Jiang subgroup has the following properties:

1.
$$I(\tilde{f}) \subset Z(\tilde{f}_*(\pi), \pi);$$

2.
$$I(id_{\tilde{X}}) \subset Z(\pi);$$

3.
$$I(\tilde{g}) \subset I(\tilde{g} \circ \tilde{f});$$

4.
$$\tilde{g}_*(I(\tilde{f})) \subset I(\tilde{g} \circ \tilde{f});$$

5.
$$I(id_{\tilde{X}}) \subset I(\tilde{f}).$$

The class of path-connected spaces X satisfying the condition $I(id_{\tilde{X}}) = \pi = \pi_1(X, x_0)$ is closed under homotopy equivalence and the topological product operation, and contains the simply connected spaces, generalized lens spaces, H-spaces and homogeneous spaces of the form G/G_0 where G is a topological group and G_0 a subgroup which is a connected, compact Lie group (for the proofs see [51]).

From theorem 27 it follows:

Theorem 38 *Suppose that $N(f^n) = R(f^n)$ for all $n > 0$, and that f is eventually commutative. Then the Nielsen zeta function is rational, and is given by*

$$N_f(z) = R_f(z) =$$

$$= \left(\left(\prod_{i=0}^{k} \det \left(1 - \Lambda^i f_{1*}^\infty \sigma z \right)^{(-1)^{i+1}} \right)^{(-1)^r} \right) * \left(\prod_{[h]} \frac{1}{1 - z^{\#[h]}} \right), \quad (3.9)$$

where σ, r, and $[h]$ are as in theorem 27. The function written here has a functional equation as described in theorem 29.

Theorem 39 *Suppose that $\tilde{f}_*(\pi) \subset I(\tilde{f})$ and $L(f^n) \neq 0$ for every $n > 0$. Then $N_f(z) = R_f(z)$ is rational and is given by (3.9). It has a functional equation as described in theorem 29. If $L(f^n) = 0$ only for finite number of n, then*

$$N_f(z) = R_f(z) \cdot \exp(P(z))$$

where $R_f(z)$ is rational and is given by (3.9) and $P(z)$ is a polynomial.

PROOF We have $\tilde{f}_*^n(\pi) \subset I(\tilde{f}^n)$ for every $n > 0$ (by property 4 and the condition $\tilde{f}_*(\pi) \subset I(\tilde{f})$). For any $\alpha \in \pi$, $p(\text{Fix }\alpha \circ \tilde{f}^n) = p(\text{Fix } \tilde{f}_*^n(\alpha) \circ \tilde{f}^n)$ by lemmas 2 and 5 and the fact that α and $\tilde{f}_*^n \alpha$ are in the same $\tilde{f}_*^n$-conjugacy class(see lemma 7). Since $\tilde{f}_*^n(\pi) \subset I(\tilde{f}^n)$, there is a homotopy $\{h_t\} : f^n \cong f^n$ which lifts to $\{\tilde{h}_t\} : \tilde{f}^n \cong \tilde{f}_*^n(\alpha) \circ \tilde{f}^n$. Hence Index $(f^n, p(\text{Fix } \tilde{f}^n)) = $ Index $(f^n, p(\text{Fix }\alpha \circ \tilde{f}^n))$. Since $\alpha \in \pi$ is arbitrary, any two fixed point classes of f^n have the same index. It immediately follows that $L(f^n) = 0$ implies $N(f^n) = 0$ and $L(f^n) \neq 0$ implies $N(f^n) = R(f^n)$. By property 1, $\tilde{f}^n(\pi) \subset I(\tilde{f}^n) \subset Z(\tilde{f}_*^n(\pi), \pi)$, so $\tilde{f}_*^n(\pi)$ is abelian. Hence $\tilde{f}_*$ is eventually commutative. If $L(f^n) \neq 0$ for every $n > 0$ then the first part of the theorem now follows from theorems 27 and 29 . If $L(f^n) = 0$ only for finite number of n,then the fraction $N_f(z)/R_f(z) = \exp(P(z))$, where $P(z)$ is a polynomial whose degree equal to maximal n, such that $L(f^n) \neq 0$. This gives the second part of the theorem.

Corollary 10 *Let the assumptions of theorem 39 hold. Then the poles and zeros of the Nielsen zeta function are complex numbers of the form $\zeta^a b$ where b is the reciprocal of an eigenvalue of one of the matrices*

$$\Lambda^i(f_{1*}^\infty) : \Lambda^i(H_1(X;\mathbb{Z})^\infty) \longrightarrow \Lambda^i(H_1(X;\mathbb{Z})^\infty) \qquad 0 \leq i \leq \text{rank } G$$

and ζ^a is a ψth root of unity where ψ is the number of periodic torsion elements in $H_1(X;\mathbb{Z})$. The multiplicities of the roots or poles $\zeta^a b$ and $\zeta^{a'} b'$ are the same if $b = b'$ and $hcf(a,\psi) = hcf(a',\psi)$.

Remark 6 *The conclusion of theorem 39 remains valid under the weaker precondition "there is an integer m such that $\tilde{f}_*^m(\pi) \subset I(\tilde{f}^m)$" instead of $\tilde{f}_*(\pi) \subset I(\tilde{f})$, but the proof is more complicated.*

From theorem 8 and results of Jiang [51] it follows:

Theorem 40 *(Trace formula for the Nielsen numbers) Suppose that there is an integer m such that $\tilde{f}_*^m(\pi) \subset I(\tilde{f}^m)$ and $L(f) \neq 0$. Then*

$$N(f) = R(f) = (-1)^{r+p} \sum_{i=0}^{k} (-1)^i \text{Tr } (\Lambda^i f_{1*}^\infty \otimes A). \qquad (3.10)$$

where k is $rgH_1(X,\mathbb{Z})^\infty$, A is linear map on the complex vector space of complex valued functions on the group $TorsH_1(X,\mathbb{Z})$, p the number of $\mu \in$ Spec f_{1*}^∞ such that $\mu < -1$, and r the number of real eigenvalues of f_{1*}^∞ whose absolute value is > 1.

Theorem 41 *Suppose that there is an integer m such that $\tilde{f}_*^m(\pi) \subset I(\tilde{f}^m)$. If $L(f^n) \neq 0$ for every $n > o$,then*

$$N_f(z) = R_f(z) = \left(\prod_{i=0}^{k} \det(1 - \Lambda^i f_{1*}^\infty \otimes A \cdot \sigma \cdot z)^{(-1)^{i+1}} \right)^{(-1)^r} \qquad (3.11)$$

If $L(f^n) = 0$ only for finite number of n, then

$$
\begin{aligned}
N_f(z) &= R_f(z) \cdot \exp P(z) \\
&= \left(\prod_{i=0}^{k} \det(1 - \Lambda^i f_{1*}^\infty \otimes A \cdot \sigma \cdot z)^{(-1)^{i+1}} \right)^{(-1)^r} \cdot \exp P(z) \quad (3.12)
\end{aligned}
$$

Where $P(z)$ is a polynomial ,A, k, p, and r are as in theorem 40.

PROOF

If $L(f^n) \neq 0$ for every $n > o$,then formula (3.11) follows from theorem 12. If $L(f^n) = 0$, then $N(f^n) = 0$. If $L(f^n) \neq 0$, then $N(f^n) = R(f^n)$(see proof of theorem 39).So the fraction $N_f(z)/R_f(z) = \exp(P(z))$, where $P(z)$ is a polynomial whose degree equal to maximal n, such that $L(f^n) \neq 0$.

Corollary 11 *Let the assumptions of theorem 41 hold. Then the poles and zeros of the Nielsen zeta function are complex numbers which are the reciprocal of an eigenvalue of one of the matrices*

$$\Lambda^i(f_{1*}^\infty \otimes A \cdot \sigma)$$

Corollary 12 *Let $I(id_{\tilde{X}}) = \pi$. If $L(f^n) \neq 0$ for all $n > 0$ then formula (3.11) is valid. If $L(f^n) = 0$ for finite number of n , then formula (3.12) is valid.*

Corollary 13 *Suppose that X is aspherical, f is eventually commutative. If $L(f^n) \neq 0$ for all $n > 0$ then formula (3.11) is valid.If $L(f^n) = 0$ for finite number of n , then formula (3.12) is valid*

3.5   Polyhedra with finite fundamental group.

For a compact polyhedron X with finite fundamental group, $\pi_1(X)$, the universal cover $\tilde{X}$ is compact, so we may explore the relation between $L(\tilde{f}^n)$ and Index $(p(\text{Fix } \tilde{f}^n))$.

Definition 13 *The number $\mu([\tilde{f}^n]) = \#\text{Fix } \tilde{f}_*^n$, defined to be the order of the finite group $\text{Fix } \tilde{f}_*^n$, is called the multiplicity of the lifting class $[\tilde{f}^n]$, or of the fixed point class $p(\text{Fix } \tilde{f}^n)$.*

Lemma 24 ([51])

$$L(\tilde{f}^n) = \mu([\tilde{f}^n]) \cdot \text{Index } (f^n, p(\text{Fix } \tilde{f}^n)).$$

Lemma 25 ([51]) *If $R(f^n) = \#\text{Coker } (1 - f_{1*}^n)$ (in particular if f is eventually commutative), then*

$$\mu([\tilde{f}^n]) = \#\text{Coker } (1 - f_{1*}^n).$$

Theorem 42 *Let X be a connected, compact polyhedron with finite fundamental group π. Suppose that the action of π on the rational homology of the universal cover $\tilde{X}$ is trivial, i.e. for every covering translation $\alpha \in \pi$, $\alpha_* = id : H_*(\tilde{X}, \mathbb{Q}) \to H_*(\tilde{X}, \mathbb{Q})$. If $L(f^n) \neq 0$ for all $n > 0$ then $N_f(z)$ is a rational function given by*

$$N_f(z) = R_f(z) = \prod_{[<h>]} \frac{1}{1 - z^{\#[<h>]}}, \tag{3.13}$$

where the product is taken over all periodic $\tilde{f}_*$-orbits of ordinary conjugacy classes in the finite group $\pi_1(X)$. This function has a functional equation as described in theorem 26. If $L(f^n) = 0$ only for finite number of n, then

$$N_f(z) = R_f(z) \cdot \exp (P(z)),$$

where $P(z)$ is a polynomial and $R_f(z)$ is given by formula (3.13)

PROOF Under our assumption on X, any two liftings $\tilde{f}^n$ and $\alpha \circ \tilde{f}^n$ induce the same homology homomorphism $H_*(\tilde{X}, \mathbb{Q}) \to H_*(\tilde{X}, \mathbb{Q})$, and have thus the same value of $L(\tilde{f}^n)$. From Lemma 24 it follows that any two fixed point classes f^n are either both essential or both inessential. If $L(f^n) \neq 0$ for every $n > 0$ then for every n there is at least one essential fixed point class of f^n. Therefore for every n all fixed point classes of f^n are essential and $N_f(z) = R_f(z)$. The formula (3.13) for $R_f(z)$ follows from theorem 16 . If $L(f^n) = 0$, then $N(f^n) = 0$. If $L(f^n) \neq 0$, then $N(f^n) = R(f^n)$. So the fraction $N_f(z)/R_f(z) = \exp(P(z))$, where $P(z)$ is a polynomial whose degree equal to maximal n, such that $L(f^n) \neq 0$. This gives the second statement of the theorem.

Let W be the complex vector space of complex valued class functions on the fundamental group π. The map $\tilde{f}_*$ induces a linear map $B : W \rightarrow W$ defined by

$$B(f) := f \circ \tilde{f}_*.$$

Theorem 43 *(Trace formula for Nielsen numbers) Let X be a connected, compact polyhedron with finite fundamental group π. Suppose that the action of π on the rational homology of the universal cover $\tilde{X}$ is trivial, i.e. for every covering translation $\alpha \in \pi$, $\alpha_* = id : H_*(\tilde{X}, \mathbb{Q}) \rightarrow H_*(\tilde{X}, \mathbb{Q})$. Let $L(f) \neq 0$. Then*

$$N(f) = R(f) = \operatorname{Tr} B, \qquad (3.14)$$

PROOF Under our assumption on X all fixed point classes of f are essential and $N(f) = R(f)$ (see proof of the previous theorem for n=1).The formula for $N(f)$ follows now from theorem 15.

Theorem 44 *Let X be a connected, compact polyhedron with finite fundamental group π. Suppose that the action of π on the rational homology of the universal cover $\tilde{X}$ is trivial, i.e. for every covering translation $\alpha \in \pi$, $\alpha_* = id : H_*(\tilde{X}, \mathbb{Q}) \rightarrow H_*(\tilde{X}, \mathbb{Q})$. If $L(f^n) \neq 0$ for every $n > o$,then*

$$N_f(z) = R_f(z) = \frac{1}{\det(1 - Bz)}, \qquad (3.15)$$

If $L(f^n) = 0$ only for finite number of n,then

$$N_f(z) = R_f(z) \cdot \exp\left(P(z)\right) = \frac{\exp\left(P(z)\right)}{\det(1 - Bz)}, \qquad (3.16)$$

Where $P(z)$ is a polynomial, B is defined in theorem 15

PROOF

If $L(f^n) \neq 0$ for every $n > o$,then $N(f^n) = R(f^n)$ (see proof of theorem 42) and formula (3.16) follows from theorem 17 . If $L(f^n) = 0$, then $N(f^n) = 0$. If $L(f^n) \neq 0$, then $N(f^n) = R(f^n)$. So the fraction $N_f(z)/R_f(z) = \exp(P(z))$, where $P(z)$ is a polynomial whose degree equal to maximal n, such that $L(f^n) \neq 0$.

Lemma 26 *Let X be a polyhedron with finite fundamental group π and let $p : \tilde{X} \to X$ be its universal covering. Then the action of π on the rational homology of $\tilde{X}$ is trivial iff $H_*(\tilde{X}; \mathbb{Q}) \cong H_*(X; \mathbb{Q})$.*

Corollary 14 *Let $\tilde{X}$ be a compact 1-connected polyhedron which is a rational homology n-sphere, where n is odd. Let π be a finite group acting freely on $\tilde{X}$ and let $X = \tilde{X}/\pi$. Then theorems 42 and 44 applie.*

PROOF The projection $p : \tilde{X} \to X = \tilde{X}/\pi$ is a universal covering space of X. For every $\alpha \in \pi$, the degree of $\alpha : \tilde{X} \to \tilde{X}$ must be 1, because $L(\alpha) = 0$ (α has no fixed points). Hence $\alpha_* = id : H_*(\tilde{X}; \mathbb{Q}) \to H_*(\tilde{X}; \mathbb{Q})$.

Corollary 15 *If X is a closed 3-manifold with finite π, then theorems 42 and 44 applie.*

PROOF $\tilde{X}$ is an orientable, simply connected manifold, hence a homology 3-sphere. We apply corollary 14 .

Corollary 16 *Let $X = L(m, q_1, \ldots, q_r)$ be a generalized lens space and $f : X \to X$ a continuous map with $f_{1*}(1) = d$ where $\mid d \mid \neq 1$. The Nielsen and Reidemeister zeta functions are then rational and are given by the formula:*

$$N_f(z) = R_f(z) = \prod_{[h]} \frac{1}{1 - z^{\#[h]}} = \prod_{t=1}^{\varphi_d(m)} (1 - e^{2\pi it/\varphi_d(m)}z)^{-a(t)}.$$

where $[h]$ runs over the periodic f_{1*}-orbits of elements of $H_1(X; \mathbb{Z})$. The numbers $a(t)$ are natural numbers given by the formula

$$a(t) = \sum_{\substack{s \mid m \text{ such that} \\ \varphi_d(m) \mid t\varphi_d(s)}} \frac{\varphi(s)}{\varphi_d(s)},$$

where φ is the Euler totient function and $\varphi_d(s)$ is the order of the multiplicative subgroup of $(\mathbb{Z}/s\mathbb{Z})^\times$ generated by d.

PROOF By corollary 14 we see that theorem 42 applies for lens spaces. Since $\pi_1(X) = \mathbb{Z}/m\mathbb{Z}$, the map f is eventually commutative. A lens space has a structure as a CW complex with one cell e_i in each dimension $0 \leq i \leq 2n + 1$. The boundary map is given by $\partial e_{2k} = m.e_{2k-1}$ for even cells, and $\partial e_{2k+1} = 0$ for odd cells. From this we may calculate the Lefschetz numbers:

$$L(f^n) = 1 - d^{(l+1)n} \neq 0.$$

This is true for any n as long as $\mid d \mid \neq 1$. Then by theorem 42 we have

$$N(f^n) = R(f^n) = \#\text{Coker } (1 - f_{1*}^n)$$

where f_{1*} is multiplication by d. One then sees that $(1 - f_{1*}^n)(\mathbb{Z}/m\mathbb{Z}) = (1 - d^n)(\mathbb{Z}/m\mathbb{Z})$ and therefore Coker $(1 - f_{1*}^n) = (\mathbb{Z}/m\mathbb{Z})/(1 - d^n)((\mathbb{Z}/m\mathbb{Z})$. The cokernel is thus a cyclic group of order $hcf(1 - d^n, m)$.

We briefly investigate the sequence $n \mapsto hcf(1 - d^n, m)$. It was originally this calculation which lead us to the results of section 2.4.

Let $\varphi : \mathbb{N} \to \mathbb{N}$ be the Euler totient function, ie. $\varphi(r) = \#(\mathbb{Z}/r\mathbb{Z})^\times$. In addition we define $\varphi_d(r)$ to be the order of the multiplicative subgroup of $(\mathbb{Z}/r\mathbb{Z})^\times$ generated by d. One then has by Lagrange's theorem $\varphi_d(r) \mid \varphi(r)$. The number $\varphi(r)$ is the smallest $n > 0$ such that $d^n \equiv 1$ mod r.

The sequence $n \mapsto hcf(1-d^n, m)$ is periodic in n with least period $\varphi_d(m)$. It can therefore be expressed as a finite Fourier series:

$$hcf(1 - d^n, m) = \sum_{t=1}^{\varphi_d(m)} a(t) \exp\left(\frac{2\pi i n t}{\varphi_d(m)}\right).$$

The coefficients $a(t)$ are given by Fourier's inversion formula:

$$a(t) = \frac{1}{\varphi_d(m)} \sum_{n=1}^{\varphi_d(m)} hcf(1 - d^n, m) \exp\left(\frac{-2\pi i n t}{\varphi_d(m)}\right).$$

After a simple calculation, one obtains the formula:

$$\exp\left(\sum_{n=1}^{\infty} \frac{hcf(1 - d^n, m)}{n} z^n\right) = \prod_{t=1}^{\varphi_d(m)} (1 - e^{2\pi i t/\varphi_d(m)} z)^{-a(t)}.$$

We now calculate the coefficients $a(t)$ more explicitly. We have $hcf(d^n - 1, m) = r$ iff $d^n \equiv 1$ mod r and for all primes $p \mid \frac{m}{r}$, $d^n \not\equiv 1$ mod pr. This is the case if and only if $n \equiv 0 \bmod \varphi_d(r)$ and for all primes $p \mid \frac{m}{r}$, $n \not\equiv 0 \bmod \varphi_d(pr)$. Using this we partition the sum in the expression for $a(t)$:

$$a(t) = \frac{1}{\varphi_d(m)} \sum_{r|m} r \sum_{\substack{n=1 \text{ such that} \\ n \equiv 0 \bmod \varphi_d(r) \text{ and} \\ \forall p \mid \frac{m}{r},\ n \not\equiv 0 \bmod \varphi_d(pr)}}^{\varphi_d(m)} \exp\left(\frac{-2\pi i n t}{\varphi_d(m)}\right).$$

We define

$$g(r) := \sum_{\substack{n=1 \text{ such that} \\ n \equiv 0 \bmod \varphi_d(r)}}^{\varphi_d(m)} \exp\left(\frac{-2\pi i n t}{\varphi_d(m)}\right).$$

Using this we rewrite the inner sum in the expression for $a(t)$.

$$\sum_{\substack{n=1 \text{ such that} \\ n \equiv 0 \bmod \varphi_d(r) \text{ and} \\ \forall p \mid \frac{m}{r},\ n \not\equiv 0 \bmod \varphi_d(pr)}}^{\varphi_d(m)} \exp\left(\frac{-2\pi i n t}{\varphi_d(m)}\right) =$$

$$= g(r) - \sum_{p_1 \mid \frac{m}{r}} g(p_1 r) + \sum_{p_1, p_2 \mid \frac{m}{r}} g(p_1 p_2 r) - \cdots$$

Here the second sum is over all 2-element sets of primes $\{p_1, p_2\}$ such that p_1 and p_2 both divide $\frac{m}{r}$. Substituting this into our expression for $a(t)$ we obtain,

$$a(t) = \frac{1}{\varphi_d(m)} \sum_{r|m} r \left(g(r) - \sum_{p_1 \mid \frac{m}{r}} g(p_1 r) + \sum_{p_1, p_2 \mid \frac{m}{r}} g(p_1 p_2 r) - \cdots \right).$$

We now change the variable of summation:

$$a(t) = \frac{1}{\varphi_d(m)} \sum_{s|m} g(s) \left(s - \sum_{p_1 | s} \frac{s}{p_1} + \sum_{p_1, p_2 | s} \frac{s}{p_1 p_2} - \cdots \right).$$

The large bracket here can be factorized into a kind of Euler product:

$$a(t) = \frac{1}{\varphi_d(m)} \sum_{s|m} g(s) \prod_{p|s} \left(1 - \frac{1}{p} \right).$$

This is then seen to be exactly the Euler totient function:

$$a(t) = \frac{1}{\varphi_d(m)} \sum_{s|m} g(s)\varphi(s).$$

On the other hand, a simple calculation shows

$$g(s) = \begin{cases} \frac{\varphi_d(m)}{\varphi_d(s)} & \text{if } \varphi_d(m) \mid t\varphi_d(s) \\ 0 & \text{otherwise.} \end{cases}$$

From this we have

$$a(t) = \sum_{\substack{s \mid m \text{ such that} \\ \varphi_d(m) \mid t\varphi_d(s)}} \frac{\varphi(s)}{\varphi_d(s)}.$$

It is now clear that the coefficients $a(t)$ are natural numbers, and we have the formula stated.

Remark 7 *Let $X = L^{2l+1}(m, q_1, q_2, \ldots, q_r)$ and let $f : X \to X$ be an continuous map. If f induces the trivial map on the cyclic group $\pi_1(X)$ then we have,*

$$N_f(z) = 1, \quad R_f(z) = \frac{1}{(1-z)^m}.$$

If f induces the map $g \mapsto -g$ on $\pi_1(X)$ then we have

$$R_f(z) = \begin{cases} \frac{1}{(1-z)^{\frac{m}{2}+1}(1+z)^{\frac{m}{2}-1}} & \text{if } m \text{ is even,} \\ \frac{1}{(1-z)^{\frac{m+1}{2}}(1+z)^{\frac{m-1}{2}}} & \text{if } m \text{ is odd,} \end{cases}$$

$$N_f(z) = \begin{cases} 1 & \text{if } l \text{ is odd,} \\ \frac{1}{1-z^2} & \text{if } l \text{ is even and } m \text{ is even,} \\ \sqrt{\frac{1+z}{1-z}} & \text{if } l \text{ is even and } m \text{ is odd.} \end{cases}$$

We have now described explicitly all the Reidemeister and Nielsen zeta functions of all continuous maps of lens spaces. Apart from one exception they are all rational.

3.6   Nielsen zeta function in other special cases

Theorems 22 and 31 implie

Theorem 45 *Let f be any continuous map of a nilmanifold M to itself.If $R(f^n)$ is finite for all n then*

$$N_f(z) = R_f(z) = \left(\prod_{i=0}^{m} \det(1 - \Lambda^i \tilde{F}.\sigma.z)^{(-1)^{i+1}} \right)^{(-1)^r} \tag{3.17}$$

where $\sigma = (-1)^p$,p , r, m and $\tilde{F}$ is defined in theorem 31.

Theorem 46 *Suppose M is a orientable compact connected 3-manifold such that $intM$ admits a complete hyperbolic structure with finite volume and $f : M \to M$ is orientation preserving homeomorphism.Then Nielsen zeta function is rational and*

$$N_f(z) = L_f(z)$$

PROOF B.Jiang and S. Wang [53] have proved that $N(f) = L(f)$. This is also true for all iterations.

3.6.1   Pseudo-Anosov homeomorphism of a compact surface

Let X be a compact surface of negative Euler characteristic and $f : X \to X$ is a pseudo-Anosov homeomorphism,i.e. there is a number $\lambda > 1$ and a pair of transverse measured foliations (F^s, μ^s) and (F^u, μ^u) such that $f(F^s, \mu^s) = (F^s, \frac{1}{\lambda}\mu^s)$ and $f(F^u, \mu^u) = (F^u, \lambda\mu^u)$. Fathi and Shub [22] has proved the existence of Markov partitions for a pseudo-Anosov homeomorphism.The existence of Markov partitions implies that there is a symbolic dynamics for (X, f).This means that there is a finite set N, a matrix $A = (a_{ij})_{(i,j)\in N\times N}$ with entries 0 or 1 and a surjective map $p : \Omega \to X$,where

$$\Omega = \{(x_n)_{n\in\mathbb{Z}} : a_{x_n x_{n+1}} = 1, n \in \mathbb{Z}\}$$

such that $p \circ \sigma = f \circ p$ where σ is the shift (to the left) of the sequence (x_n) of symbols.We have first [11]:

$$\#\text{Fix } \sigma^n = \text{Tr } A^n.$$

In general p is not bijective. The non-injectivity of p is due to the fact that the rectangles of the Markov partition can meet on their boundaries. To cancel the overcounting of periodic points on these boundaries, we use Manning's combinatorial arguments [64] proposed in the case of Axiom A diffeomorphism (see also [72]) . Namely, we construct finitely many subshifts of finite type $\sigma_i, i = 0, 1, .., m$, such that $\sigma_0 = \sigma$, the other shifts semi-conjugate with restrictions of f [72] ,and signs $\epsilon_i \in \{-1, 1\}$ such that for each n

$$\#\text{Fix } f^n = \sum_{i=0}^{m} \epsilon_i \cdot \#\text{Fix } \sigma_i^n = \sum_{i=0}^{m} \epsilon_i \cdot \text{Tr } A_i^n,$$

where A_i is transition matrix, corresponding to subshift of finite type σ_i. For pseudo-Anosov homeomorphism of compact surface $N(f^n) = \#\text{Fix } (f^n)$ for each $n > o$ [91]. So we have following trace formula for Nielsen numbers

Lemma 27 *Let X be a compact surface of negative euler characteristic and $f : X \to X$ is a pseudo-Anosov homeomorphism. Then*

$$N(f^n) = \sum_{i=0}^{m} \epsilon_i \cdot \text{Tr } A_i^n.$$

This lemma implies

Theorem 47 *Let X be a compact surface of negative Euler characteristic and $f : X \to X$ is a pseudo-Anosov homeomorphism. Then*

$$N_f(z) = \prod_{i=0}^{m} \det(1 - A_i z)^{-\epsilon_i} \qquad (3.18)$$

where A_i and ϵ_i the same as in lemma 27.

3.7 The Nielsen zeta function and Serre bundles.

Let $p : E \to B$ be a orientable Serre bundle in which E, B and every fibre are connected, compact polyhedra and $F_b = p^{-1}(b)$ is a fibre over $b \in B$ (see section). Let $f : E \to E$ be a fibre map. Then for any two fixed points b, b'

of $\bar{f} : B \to B$ the maps $f_b = f \mid_{F_b}$ and $f_{b'} = f \mid_{F_{b'}}$ have the same homotopy type; hence they have the same Nielsen numbers $N(f_b) = N(f_{b'})$. The following theorem describes the relation between the Nielsen zeta functions $N_f(z)$, $N_{\bar{f}}(z)$ and $N_{f_b}(z)$ for a fibre map $f : E \to E$ of an orientable Serre bundle $p : E \to B$.

Theorem 48 *Suppose that for every $n > 0$*

   1) $KN(f_b^n) = N(f_b)$, where $b \in$ Fix $(\bar{f}^n), K = K_b = Ker(i_* : \pi_1(F_b) \to \pi_1(E))$;

   2) in every essential fixed point class of f^n, there is a point e such that

$$p_*(\text{Fix } (\pi_1(E, e) \overset{(f^n)*}{\to} \pi_1(E, e) == \text{Fix } (\pi_1(B, b_0) \overset{(\bar{f}^n)*}{\to} \pi_1(B, b_0),$$

where $b_0 = p(e)$. We then have

$$N_f(z) = N_{\bar{f}}(z) * N_{f_b}(z).$$

If $N_{\bar{f}}(z)$ and $N_{f_b}(z)$ are rational functions then so is $N_f(z)$. If $N_{\bar{f}}(z)$ and $N_{f_b}(z)$ are rational functions with functional equations as described in theorem 38 and 42 then so is $N_f(z)$.

PROOF From the conditions of the theorem it follows that

$$N(f^n) = N(\bar{f}^n) \cdot N(f_b^n)$$

for every $n($ see [51]). From this we have

$$N_f(z) = N_{\bar{f}}(z) * N_{f_b}(z).$$

The rationality of $N_f(z)$ and functional equation for it follow from lemmas 13 and 14 .

Corollary 17 *Suppose that for every $n > 0$ homomorphism $1 - (\bar{f}^n)_* : \pi_2(B, b) \to \pi_2(B, b)$ is an epimorphism. Then the condition 1) above is satisfied.*

Corollary 18 *Suppose that $f : E \to E$ admits a Fadell splitting in the sense that for some e in Fix f and $b = p(e)$ the following conditions are satisfied:*

1. the sequence

$$0 \longrightarrow \pi_1(F_b, e) \xrightarrow{\ i_* \ } \pi_1(E, e) \xrightarrow{\ p_* \ } \pi_1(B, e) \longrightarrow 0$$

is exact,

2. p_* admits a right inverse (section) σ such that Im σ is a normal subgroup of $\pi_1(E, e)$ and $f_*(\text{Im } \sigma) \subset \text{Im } \sigma$.

Then theorem 48 applies.

3.8   Examples

Let $f : X \to X$ be a continuous map of a simply connected, connected, compact polyhedron. Then $R_f(z) = \frac{1}{1-z}$.

Let $X = S^1$ and $f : S^1 \to S^1$ be continuous map of degree d. Then $N(f^n) = | 1 - d^n |$, and the Nielsen zeta function is rational and is equal to

$$N_f(z) = \begin{cases} \frac{1-z}{1-dz} & \text{if } d > 0 \\ \frac{1}{1-z} & \text{if } d = 0 \\ \frac{1+z}{1+dz} & \text{if } d < 0 \ . \end{cases}$$

If $X = S^{2n}$, and $f : S^{2n} \to S^{2n}$ is a continuous map of degree d then

$$N_f(z) = \begin{cases} \frac{1}{\sqrt{1-z^2}} & \text{if } d = -1 \\ \frac{1}{1-z} & \text{if } d \neq -1. \end{cases}$$

Now if $X = S^{2n+1}$, and $f : S^{2n+1} \to S^{2n+1}$ is a continuous map of degree d then

$$N_f(z) = \begin{cases} 1 & \text{if } d = 1 \\ \sqrt{\frac{1+z}{1-z}} & \text{if } d = -1 \\ \frac{1}{1-z} & \text{if } | d | \neq 1. \end{cases}$$

Thus, even on a simply connected space the Nielsen zeta function can be the radical of a rational function. In the next example $X = T^n$ is torus and $f : T^n \to T^n$ is a hyperbolic endomorphism of the torus . Hyperbolic means that the covering linear map $\tilde{f} : R^n \to R^n$ has no eigenvalues of modulus one. Then $R(f^n) = N(f^n) = |\det(E - \tilde{f}^n)| = |L(f^n)|$ [12]. Thus $R[f^n] = N(f^n) = (-1)^{r+pn} \cdot \det(E - \tilde{f}^n)$, where r is equal to the number of $\lambda_i \in Spec(\tilde{f})$ such that $|\lambda_i| > 1$, and p is equal to the number of $\mu_i \in Spec(\tilde{f})$ such that $\mu_i < -1$. Consequently, $R[f^n] = N(f^n) = (-1)^{r+pn} \cdot L(f^n)$ and the Reidemeister and Nielsen zeta function are rational and equal to $R_f(z) = N_f(z) = (L_f(\sigma \cdot z))^{(-1)^r}$, where $\sigma = (-1)^p$. It follows from the results of Franks, Newhouse and Manning [85] that the following diffeomorphisms g are topologically conjugate to hyperbolic automorphisms of the torus Γ: a Anosov diffeomorphism of the torus, a Anosov diffeomorphism of codimension one [85] of manifold , which is metrically decomposable [85] , a Anosov diffeomorphism of a manifold , whose fundamental group is commutative. Consequently, by the topological conjugacy of g and Γ, the Nielsen zeta function $N_g(z)$ is rational and equal to $N_g(z) = N_\Gamma(z) = (L_f(\sigma \cdot z))^{(-1)^r}$. In this example the Reidemeister and Nielsen zeta functions coincide with the Artin-Mazur zeta function. In fact, the covering map $\tilde{\Gamma}$ has a unique fixed point, which is the origin; hence, by the covering homotopy theorem [80] , the fixed points of Γ are pairwise nonequivalent. The index of each equivalence class, consisting of one fixed point, coincides with its Lefschetz index, and by the hyperbolicity of Γ, the later is not equal to zero. Thus $R(\Gamma) = N(\Gamma) = F(\Gamma)$. Analogously, $R(\Gamma^n) = N(\Gamma^n) = F(\Gamma^n)$ for each $n > o$ and $R_\Gamma(z) = N_\Gamma(z) = F_\Gamma(z)$. Since Γ satisfies axiom A of Smale, by Manning theorem [64] we get another proof of the rationality of $R_\Gamma(z) = N_\Gamma(z) = F_\Gamma(z)$.

let $X = RP^{2k+1}, k \neq 0$ be projective space of odd dimension. Then for each $n > 0, N(f^n) = 0$, if $d = 1$, and $N(f^n) = (2, 1 - d^n)$, if $|d| \neq 1$. Consequently, $N(f^n) = 2$, for all $n > 0$, if $|d| \neq 1$ is odd , and $N(f^n) = 1$ for all $n > 0$, if d is even, and the Nielsen zeta function is rational and equal to:

$$N_f(z) = \begin{cases} 1 & \text{if } d = 1 \\ \frac{1}{(1-z)^2} & \text{if } |d| \neq 1 \text{ is odd} \\ \frac{1}{1-z} & \text{if } d \text{ is even.} \end{cases}$$

Now if $d = -1$, then for even $n, N(f^n) = 0$, for odd $n, N(f^n) = 2$, and

$N_f(z) = (1+z)/(1-z)$. For RP^{2k}, the projective spaces of even dimension, one gets exactly the same result for $N_f(z)$.

Now let $f : M \to M$ be an expanding map[86] of the orientable smooth compact manifold M. Then M is aspherical and is $K(\pi_1(M), 1)$ and the fundamental group $\pi_1(M)$ is torsion free [86] . If $pr : \tilde{M} \to M$ is the universal covering, $\tilde{f} : \tilde{M} \to \tilde{M}$ is an arbitrary map , covering f, then according to Shub [86] $\tilde{f}$ has exactly one fixed point. From this and the covering homotopy theorem [80] it follows that the fixed points of f are pairwise nonequivalent. The index of each equivalence class, consisting of one fixed point, coincides with its Lefschetz index,.If f preserves the orientation of M, then the Lefschetz index $L(p, f)$ of the fixed point p is equal to $L(p, f) = (-1)^r$, where $r = \dim M$. Then by Lefschetz trace formula $R(f) = N(f) = F(f) = (-1)^r \cdot L(f)$.Since the iterates f^n are also orientation preserving expanding maps, we get analogously that $R(f^n) = N(f^n) = F(f^n) = (-1)^r \cdot L(f^n)$ for every n, and the Nielsen zeta function is rational and equal to $R_f(z) = N_f(z) = F_f(z) = L_f(z)^{(-1)^r}$. Now if f reverses the orientation of the manifold M, then $L(p, f^n) = (-1)^{r+n}$. Hence , $R(f^n) = N(f^n) = F(f^n) = (-1)^{r+n} \cdot L(f^n)$ and $R_f(z) = N_f(z) = F_f(z) = L_f(-z)^{(-1)^r}$.

Example 8 ([8]) *Let $f : S^2 \vee S^4 \to S^2 \vee S^4$ to be a continuous map of the bouquet of spheres such that the restriction $f/S^4 = id_{S^4}$ and the degree of the restriction $f/S^2 : S^2 \to S^2$ equal to -2.Then $L(f) = 0$, hence $N(f) = 0$ since $S^2 \vee S^4$ is simply connected.For $k > 1$ we have $L(f^k) = 2 + (-2)^k \neq 0$,therefore $N(f^k) = 1$.From this we have by direct calculation that*

$$N_f(z) = \exp(-z) \cdot \frac{1}{1-z}. \tag{3.19}$$

Remark 8 *We would like to mention that in all known cases the Nielsen zeta function is a nice function. By this we mean that it is a product of an exponential of a polynomial with a function some power of which is rational. May be this is a general pattern; it could however be argued that this just reflects our inability to calculate the Nielsen numbers in general case.*

Chapter 4

Reidemeister and Nielsen zeta functions modulo normal subgroup, minimal dynamical zeta functions

4.1 Reidemeister and Nielsen zeta functions modulo a normal subgroup

In the theory of (ordinary) fixed point classes, we work on the universal covering space. The group of covering transformations plays a key role. It is not surprising that this theory can be generalized to work on all regular covering spaces. Let K be a normal subgroup of the fundamental group $\pi_1(X)$. Consider the regular covering $p_K : \tilde{X}/K \to X$ corresponding to K. A map $\tilde{f}_K : \tilde{X}/K \to \tilde{X}/K$ is called a lifting of $f : X \to X$ if $p_K \circ \tilde{f}_K = f \circ p_K$. We know from the theory of covering spaces that such liftings exist if and only if $f_*(K) \subset K$. If K is a fully invariant subgroup of $\pi_1(X)$ (in the sense that every endomorphism sends K into K) such as, for example the commutator subgroup of $\pi_1(X)$, then there is a lifting $\tilde{f}_K$ of any continuous map f.

We can develop a theory which is similar to the theory in Chapters I - II by simply replacing $\tilde{X}$ and $\pi_1(X)$ by $\tilde{X}/K$ and $\pi_1(X)/K$ in every definition, every theorem and every proof, since everything was done in terms of liftings and covering translations. What follows is a list of definitions and some basic facts.

Two liftings $\tilde{f}_K$ and $\tilde{f}'_K$ are called *conjugate* if there is a $\gamma_K \in \Gamma_K \cong \pi_1(X)/K$ such that $\tilde{f}'_K = \gamma_K \circ \tilde{f}_K \circ \gamma_K^{-1}$. The subset $p_K(\text{Fix } (\tilde{f}_K)) \subset \text{Fix } (f)$ is called the mod K fixed point class of f determined by the lifting class $[\tilde{f}_K]$ on $\tilde{X}/K$. The fixed point set Fix (f) splits into a disjoint union of mod K fixed point classes. Two fixed points x_0 and x_1 belong to the same mod K class iff there is a path c from x_0 to x_1 such that $c * (f \circ c)^{-1} \in K$. Each mod K fixed point class is a disjoint union of ordinary fixed point classes. So the index of a mod K fixed point class can be defined in obvious way. A mod K fixed point class is called *essential* if its index is nonzero. The number of lifting classes of f on $\tilde{X}/K$ (and hence the number of mod K fixed point classes, empty or not) is called the mod K Reidemeister Number of f, denoted $KR(f)$. This is a positive integer or infinity. The number of essential mod K fixed point classes is called the mod K Nielsen number of f, denoted by $KN(f)$. The mod K Nielsen number is always finite. $KR(f)$ and $KN(f)$ are homotopy type invariants. The mod K Nielsen number was introduced by G.Hirsch in 1940 , primarily for purpose of estimating Nielsen number from below. The mod K Reidemeister zeta functions of f and the mod K Nielsen zeta function of f was defined in [24], [26] as power series:

$$KR_f(z) := \exp\left(\sum_{n=1}^{\infty} \frac{KR(f^n)}{n} z^n\right),$$

$$KN_f(z) := \exp\left(\sum_{n=1}^{\infty} \frac{KN(f^n)}{n} z^n\right).$$

$KR_f(z)$ and $KN_f(z)$ are homotopy invariants. If K is the trivial subgroup of $\pi_1(X)$ then $KN_f(z)$ and $KR_f(z)$ coincide with the Nielsen and Reidemeister zeta functions respectively.

4.1.1 Radius of Convergence of the mod K Nielsen zeta function

We show that the mod K Nielsen zeta function has positive radius of convergence.We denote by R the radius of convergence of the mod K Nielsen zeta function $N_f(z)$,

Theorem 49 *Suppose that* $f : X \to X$ *be a continuous map of a compact polyhedron and* $f_*(K) \subset K$. *Then*

$$R \geq \exp(-h) > 0 \qquad (4.1)$$

and

$$R \geq \frac{1}{\max_d \|z\tilde{F}_d\|} > 0, \qquad (4.2)$$

and

$$R \geq \frac{1}{\max_d s(\tilde{F}_d^{norm})} > 0, \qquad (4.3)$$

where $\tilde{F}_d$ *and* h *is the same as in section 3.1.1 and 3.1.2.*

PROOF The theorem follows from inequality $N(f^n) \geq KN(f^n)$, Caushy-Adamar formula and theorems 33 and 34 .

Remark 9 *Let* f *be a* C^1-*mapping of a compact, smooth, Riemannian manifold* M. *Then* $h(f) \leq \log \sup \| Df(x)* \|$ *[83], where* $Df(x)*$ *is a mapping between exterior algebras of the tangent spaces* $T(x)$ *and* $T(f(x))$, *induced by* $Df(x)$, $\| \cdot \|$ *is the norm on operators, induced from the Riemann metric. Now from the inequality*

$$h(f) \geq \limsup_n \frac{1}{n} \cdot \log KN(f^n)$$

and the Cauchy-Adamar formula we have

$$R \geq \frac{1}{\sup_{x \in M} \|D(f) * (x)\|}, \qquad (4.4)$$

4.1.2   mod K Nielsen zeta function of a periodic map

We denote $KN(f^n)$ by KN_n.Let $\mu(d), d \in N$, be the Möbius function

Theorem 50 *Let* f *be a periodic map of least period* m *of the connected compact polyhedron* X *and* $f_*(K) \subset K$. *Then the mod K Nielsen zeta function is equal to*

$$KN_f(z) = \prod_{d|m} \sqrt[d]{(1 - z^d)^{-P(d)}},$$

*where the product is taken over all divisors d of the period m, and $P(d)$ is
the integer*

$$P(d) = \sum_{d_1|d} \mu(d_1) KN_{d|d_1}.$$

PROOF Since $f^m = id$, for each $j, KN_j = KN_{m+j}$. Since $(k,m) = 1$, there
exist positive integers t and q such that $kt = mq + 1$. So $(f^k)^t = f^{kt} = f^{mq+1} = f^{mq}f = (f^m)^q f = f$. Consequently, $KN((f^k)^t) = KN(f)$. Let two
fixed point x_0 and x_1 belong to the same mod K fixed point class. Then
there exists a path α from x_0 to x_1 such that $\alpha * (f \circ \alpha)^{-1} \in K$.Since
$f_*(K) \subset K$, we have $f(\alpha * f \circ \alpha)^{-1}) = (f \circ \alpha) * (f^2 \circ \alpha)^{-1} \in K$ and a product
$\alpha * (f \circ \alpha)^{-1} * (f \circ \alpha) * (f^2 \circ \alpha)^{-1} = \alpha * (f^2 \circ \alpha)^{-1} \in K$. It follows that
$\alpha * (f^k \circ \alpha)^{-1} \in K$ is derived by the iteration of this process. So x_0 and x_1
belong to the same mod K fixed point class of f^k. If two point belong to the
different mod K fixed point classes f, then they belong to the different mod
K fixed point classes of f^k.So, each essential class(class with nonzero index)
for f is an essential class for f^k; in addition , different essential classes for
f are different essential classes for f^k. So $KN(f^k) \geq KN(f)$. Analogously,
$KN(f) = KN((f^k)^t) \geq KN(f^k)$.Consequently , $KN(f) = KN(f^k)$. One
can prove completely analogously that $KN_d = KN_{di}$, if (i, m/d) =1, where
d is a divisor of m. Using these series of equal mod K Nielsen numbers, one
can regroup the terms of the series in the exponential of the mod K Nielsen
zeta function so as to get logarithmic functions by adding and subtracting
missing terms with necessary coefficient:

$$
\begin{aligned}
KN_f(z) &= \exp\left(\sum_{i=1}^{\infty} \frac{KN(f^i)}{i} z^i\right) \\
&= \exp\left(\sum_{d|m}\sum_{i=1}^{\infty} \frac{P(d)}{d} \cdot \frac{z^{di}}{i}\right) \\
&= \exp\left(\sum_{d|m} \frac{P(d)}{d} \cdot \log(1 - z^d)\right) \\
&= \prod_{d|m} \sqrt[d]{(1 - z^d)^{-P(d)}}
\end{aligned}
$$

where the integers $P(d)$ are calculated recursively by the formula

$$P(d) = KN_d - \sum_{d_1|d;d_1\neq d} P(d_1).$$

Moreover, if the last formula is rewritten in the form

$$KN_d = \sum_{d_1 | d} \mu(d_1) \cdot P(d_1)$$

and one uses the Möbius Inversion law for real function in number theory, then

$$P(d) = \sum_{d_1 | d} \mu(d_1) \cdot KN_{d/d_1},$$

where $\mu(d_1)$ is the Möbius function. The theorem is proved.

4.2   Minimal dynamical zeta function

4.2.1   Radius of Convergence of the minimal zeta function

In the Nielsen theory for periodic points, it is well known that $N(f^n)$ is sometime poor as a lower bound for the number of fixed points of f^n. A good homotopy invariant lower bound $NF_n(f)$,called the Nielsen type number for f^n,is defined in [51].Consider any finite set of periodic orbit classes $\{O^{k_j}\}$ of varied period k_j such that every essential periodic m-orbit class, $m|n$, contains at least one class in the set.Then $NF_n(f)$ is the minimal sum $\sum_j k_j$ for all such finite sets. Halpern (see [51]) has proved that for all n $NF_n(f) = \min\{\#\text{Fix}\,(g^n)|g$ has the same homotopy type as f $\}$.Recently, Jiang [52] found that as far as asymptotic growth rate is concerned,these Nielsen type numbers are no better than the Nielsen numbers.

Lemma 28 ([52])

$$\limsup_n (N(f^n))^{\frac{1}{n}} = \limsup_n (NF_n(f))^{\frac{1}{n}} \qquad (4.5)$$

We define minimal dynamical zeta function as power series

$$M_f(z) := \exp\left(\sum_{n=1}^{\infty} \frac{NF_n(f)}{n} z^n\right),$$

Theorem 51 *For any continuous map f of any compact polyhedron X into itself the minimal zeta function has positive radius of convergence R, which admits following estimations*

$$R \geq \exp(-h) > 0, \tag{4.6}$$

$$R \geq \frac{1}{\max_d \|z\tilde{F}_d\|} > 0, \tag{4.7}$$

and

$$R \geq \frac{1}{\max_d s(\tilde{F}_d^{norm})} > 0, \tag{4.8}$$

where $\tilde{F}_d$ and h is the same as in section 3.1.1 and 3.1.2.

PROOF The theorem follows from Caushy-Adamar formula, lemma 28 and theorems 33 and 34.

Remark 10 *Let us consider a smooth compact manifold M, which is a regular neighborhood of X and a smooth map $g : M \to M$ of the same homotopy type as f. There is a smooth map $\phi : M \to M$ homotopic to g such that for every n iteration ϕ^n has only a finite number of fixed points $F(\phi^n)$ (see [51], p.62). According to Artin and Mazur [5] there exists constant $c = c(\phi) < \infty$, such that $F(\phi^n) < c^n$ for every $n > 0$. Then due to Halperin result $c^n > F(\phi^n) \geq NF_n(f)$ for every $n > 0$. Now the Cauchy-Adamar formula gives us the second proof that the radius of the convergence R is positive.*

Chapter 5

Congruences for Reidemeister and Nielsen numbers

5.1 Irreducible Representation and the Unitary Dual of G

Let V be a Hilbert space. A unitary representation of G on V is a homomorphism $\rho : G \to \mathrm{U}(V)$ where $\mathrm{U}(V)$ is the group of unitary transformations of V. Two of these $\rho_1 : G \to \mathrm{U}(V_1)$ and $\rho_2 : G \to \mathrm{U}(V_2)$ are said to be equivalent if there is a Hilbert space isomorphism $V_1 \cong V_2$ which commutes with the G-actions. A representation $\rho : G \to \mathrm{U}(V)$ is said to be irreducible if there is no decomposition

$$V \cong V_1 \oplus V_2$$

in which V_1 and V_2 are non-zero, closed G-submodules of V.

One defines the unitary dual $\hat{G}$ of G to be the set of all equivalence classes of irreducible, unitary representations of G. There is a multivalued map $\hat{\phi} : \hat{G} \to \hat{G}$ defined as follows. If $\rho : G \to \mathrm{U}(V)$ is an irreducible unitary representation then $\rho \circ \phi : G \to \mathrm{U}(V)$ is also a representation of G on V, which we shall denote $\phi^*(\rho)$. The representation $\phi^*(\rho)$ is a sum of irreducible pieces; we define $\hat{\phi}(\rho)$ to be this set of pieces. If ϕ is bijective or G is Abelian then $\hat{\phi}$ is single valued.

Definition 14 *Define the number $\#\mathrm{Fix}(\hat{\phi})$ to be the number of fixed points of the induced map $\hat{\phi} : \hat{G} \to \hat{G}$. We shall write $\mathcal{S}(\phi)$ for the set of fixed points* *of $\hat{\phi}$. Thus $\mathcal{S}(\phi)$ is the set of equivalence classes of irreducible representations $\rho : G \to U(V)$ such that there is a transformation $M \in U(V)$ satisfying*

$$\forall x \in G, \quad \rho(\phi(x)) = M \cdot \rho(x) \cdot M^{-1}. \tag{5.1}$$

Note that if ϕ is an inner automorphism $x \mapsto gxg^{-1}$ then we have for any representation ρ,

$$\rho(\phi(x)) = \rho(g) \cdot \rho(x) \cdot \rho(g)^{-1},$$

implying that the class of ρ is fixed by the induced map. Thus for an inner automorphism the induced map is trivial and $\#\mathrm{Fix}(\hat{\phi})$ is the cardinality of $\hat{G}$. When G is Abelian the group $\hat{G}$ is the Pontryagin dual of G.

5.2 Endomorphism of the Direct Sum of a Free Abelian and a Finite Group

In this section let F be a finite group and k a natural number. The group G will be the direct sum

$$G = \mathbb{Z}^k \oplus F$$

We shall describe the Reidemeister numbers of endomorphism $\phi : G \to G$. The torsion elements of G are exactly the elements of the finite, normal subgroup F. For this reason we have $\phi(F) \subset F$. Let $\phi^{finite} : F \to F$ be the restriction of ϕ to F, and let $\phi^{\infty} : G/F \to G/F$ be the induced map on the quotient group. We have proved in proposition 3 that

$$R(\phi) = R(\phi^{finite}) \times R(\phi^{\infty}).$$

We shall prove the following result:

Proposition 4 *In the notation described above*

$$\#\mathrm{Fix}\,(\hat{\phi}) = \#\mathrm{Fix}\,(\hat{\phi}^{finite}) \times \#\mathrm{Fix}\,(\hat{\phi}^{\infty})$$

Proof
    Consider the dual $\hat{G}$. This is cartesian product of the duals of $\mathbb{Z}^k$ and F:

$$\hat{G} = \hat{\mathbb{Z}}^k \times \hat{F}, \; \rho = \rho_1 \otimes \rho_2$$

where ρ_1 is an irreducible representation of $\mathbb{Z}^k$ and ρ_2 is an irreducible representation of F. Since $\mathbb{Z}^k$ is abelian, all of its irreducible representations are 1-dimensional , so $\rho(v)$ for $v \in \mathbb{Z}^r$ is always a scalar matrix, and ρ_2 is the restriction of rho to F. If $\rho = \rho_1 \otimes \rho_2 \in \mathcal{S}(\phi)$ then there is a matrix T such that

$$\rho \circ \phi = T \cdot \rho \cdot T^{-1}.$$

This implies

$$\rho^{finite} \circ \phi^{finite} = T \cdot \rho^{finite} \cdot T^{-1},$$

so $\rho_2 = \rho^{finite}$ is in $\mathcal{S}(\phi^{finite})$. For any fixed $\rho_2 \in \mathcal{S}(\phi^{finite})$, the set of ρ_1 with $\rho_1 \otimes \rho_2 \in \mathcal{S}(\phi^{finite})$ is the set of ρ_1 satisfying

$$\rho_1(M \cdot v)\rho_2(\psi(v)) = T \cdot \rho_1(v) \cdot T^{-1}$$

for some matrix T independent of $v \in \mathbb{Z}^r$. Since $\rho_1(v)$ is a scalar matrix, the equation is equivalent to

$$\rho_1(M \cdot v)\rho_2(\psi(v)) = \rho_1(v),$$

i.e.

$$\rho_1((1 - M)v) = \rho_2(\psi(v)).$$

Note that $\hat{\mathbb{Z}}^k$ is isomorphic to the torus T^k, and the transformation $\rho_1 \to \rho_1 \circ (1 - M)$ is given by the action of the matrix $1 - M$ on the torus T^k. Therefore the number of ρ_1 satisfying the last equation is the degree of the map $(1 - M)$ on the torus , i.e. $| \det(1 - M) |$. From this it follows that

$$\#\text{Fix } (\hat{\phi}) = \#\text{Fix } (\hat{\phi}^{finite}) \times | \det(1 - M) | .$$

As in the proof of proposition 3 we have $R(\phi^\infty) = | \det(1 - M) |$ .Since ϕ^∞ is an endomorphism of an abelian group we have $\#\text{Fix } (\hat{\phi}^\infty) = R(\phi^\infty)$.Therefore

$$\#\text{Fix } (\hat{\phi}) = \#\text{Fix } (\hat{\phi}^{finite}) \times \#\text{Fix } (\hat{\phi}^\infty).$$

As a consequence we have the following

Theorem 52 *If ϕ be any endomorphism of G where G is the direct sum of a finite group F with a finitely generated free Abelian group, then*

$$R(\phi) = \#\text{Fix } (\hat{\phi})$$

PROOF Let ϕ^{finite} is an endomorphism of a finite group F and V be the complex vector space of class functions on the group F. A class function is a function which takes the same value on every element of a (usual) congruence class. The map ϕ^{finite} induces a map

$$\varphi : V \to V$$
$$f \mapsto f \circ \phi^{finite}$$

We shall calculate the trace of φ in two ways. The characteristic functions of the congruence classes in F form a basis of V, and are mapped to one another by φ (the map need not be a bijection). Therefore the trace of φ is the number of elements of this basis which are fixed by φ. By Theorem 14 , this is equal to the Reidemeister number.

Another basis of V, which is also mapped to itself by φ is the set of traces of irreducible representations of F (see [60] chapter XVIII). From this it follows that the trace of φ is the number of irreducible representations ρ of F such that ρ has the same trace as $\hat{\phi}^{finite}(\rho)$. However, representations of finite groups are characterized up to equivalence by their traces. Therefore the trace of φ is equal to the number of fixed points of $\hat{\phi}^{finite}$.

So, we have $R(\phi^{finite}) = \#\text{Fix } (\hat{\phi}^{finite})$. Since ϕ^{∞} is an endomorphism of the finitely generated free Abelian group we have $R(\phi^{\infty}) = \#\text{Fix } (\hat{\phi}^{\infty})$ (see formula (2.13)). It now follows from propositions 3 and 4 that $R(\phi) = \#\text{Fix } (\hat{\phi})$.

Remark 11 *By specialising to the case when G is finite and ϕ is the identity map, we obtain the classical result equating the number of irreducible representation of a finite group with the number of conjugacy classes of the group.*

5.3 Endomorphism of almost Abelian groups

In this section let G be an almost Abelian and finitely generated group.A group will be called almost Abelian if it has an Abelian subgroup of finite index. We shall prove in this section an analog of theorem 52 for almost

Abelian group founded by Richard Hill [50]. It seems plausible that one could prove the same theorem for the so - called "tame" topological groups (see [57]). However we shall be interested mainly in discrete groups, and it is known that the discrete tame groups are almost Abelian.

We shall introduce the profinite completion $\overline{G}$ of G and the corresponding endomorphism $\overline{\phi} : \overline{G} \to \overline{G}$. This is a compact totally disconnected group in which G is densely embedded. The proof will then follow in three steps:

$$R(\phi) = R(\overline{\phi}), \quad \#\text{Fix } (\hat{\phi}) = \#\text{Fix } (\hat{\overline{\phi}}), \quad R(\overline{\phi}) = \#\text{Fix } (\hat{\overline{\phi}}).$$

If one omits the requirement that G is almost Abelian then one can still show that $R(\phi) \geq R(\overline{\phi})$ and $\#\text{Fix } (\hat{\phi}) \geq \#\text{Fix } (\hat{\overline{\phi}})$. The third identity is a general fact for compact groups (Theorem 53).

5.3.1   Compact Groups

Here we shall prove the third of the above identities.

Let K be a compact topological group and ϕ a continuous endomorphism of K. We define the number $\#\text{Fix }^{\text{top}}(\hat{\phi})$ to be the number of fixed points of $\hat{\phi}$ in the unitary dual of K, where we only consider continuous representations of K. The number $R(\phi)$ is defined as usual.

Theorem 53 ([50]) *For a continuous endomorphism ϕ of a compact group K one has $R(\phi) = \#\text{Fix }^{\text{top}}(\hat{\phi})$.*

The proof uses the Peter-Weyl Theorem:

Theorem 54 (Peter - Weyl) *If K is compact then there is the following decomposition of the space $L^2(K)$ as a $K \oplus K$-module.*

$$L^2(K) \cong \bigoplus_{\lambda \in \hat{K}} \text{Hom}_C(V_\lambda, V_\lambda).$$

and Schur's Lemma:

Lemma 29 (Schur) *If V and W are two irreducible unitary representations then*

$$\text{Hom}_{CK}(V, W) \cong \begin{cases} 0 & V \not\cong W \\ C & V \cong W. \end{cases}$$

PROOF OF THEOREM 53. The ϕ-conjugacy classes, being orbits of a compact group, are compact. Since there are only finitely many of them, they are also open subsets of K and thus have positive Haar measure.

We embed K in $K \oplus K$ by the map $g \mapsto (g, \phi(g))$. This makes $L^2(K)$ a K-module with a twisted action. By the Peter-Weyl Theorem we have (as K-modules)

$$L^2(K) \cong \bigoplus_{\lambda \in \hat{K}} \text{Hom}_C(V_\lambda, V_{\hat{\phi}(\lambda)}).$$

We therefore have a corresponding decomposition of the space of K-invariant elements:

$$L^2(K)^K \cong \bigoplus_{\lambda \in \hat{K}} \text{Hom}_{CK}(V_\lambda, V_{\hat{\phi}(\lambda)}).$$

We have used the well known identity $\text{Hom}_C(V, W)^K = \text{Hom}_{CK}(V, W)$.

The left hand side consists of functions $f : K \to C$ satisfying $f(gx\phi(g)^{-1}) = f(x)$ for all $x, g \in K$. These are just functions on the ϕ-conjugacy classes. The dimension of the left hand side is thus $R(\phi)$. On the other hand by Schur's Lemma the dimension of the right hand side is $\#\text{Fix}^{\text{top}}(\hat{\phi})$.

5.3.2 Almost Abelian groups

Let G be an almost Abelian group with an Abelian subgroup A of finite index $[G : A]$. Let A^0 be the intersection of all subgroups of G of index $[G : A]$. Then A^0 is an Abelian normal subgroup of finite index in G and one has $\phi(A^0) \subset A^0$ for every endomorphism ϕ of G.

Lemma 30 *If $R(\phi)$ is finite then so is $R(\phi|_{A^0})$.*

PROOF. A ϕ-conjugacy class is an orbit of the group G. A $\phi|_{A^0}$-conjugacy class is an orbit of the group A^0. Since A^0 has finite index in G it follows that every ϕ-conjugacy class in A^0 can be the union of at most finitely many $\phi|_{A^0}$-conjugacy classes. This proves the lemma.

Let $\overline{G}$ be the profinite completion of G with respect to its normal subgroups of finite index. There is a canonical injection $G \to \overline{G}$ and the map ϕ can be extended to a continuous endomorphism $\overline{\phi}$ of $\overline{G}$.

There is therefore a canonical map

$$\mathcal{R}(\phi) \to \mathcal{R}(\overline{\phi}).$$

Since G is dense in $\overline{G}$, the image of a ϕ-conjugacy class $\{x\}_\phi$ is its closure in $\overline{G}$. From this it follows that the above map is surjective. We shall actually see that the map is bijective. This will then give us

$$R(\phi) = R(\overline{\phi}).$$

However $\overline{\phi}$ is an endomorphism of the compact group $\overline{G}$ so by Theorem 53

$$R(\overline{\phi}) = \#\text{Fix}^{\text{top}}(\hat{\overline{\phi}}).$$

It thus suffices to prove the following two lemmas:

Lemma 31 *If $R(\phi)$ is finite then $\#\text{Fix}^{\text{top}}(\hat{\overline{\phi}}) = \#\text{Fix}(\hat{\phi})$.*

Lemma 32 *If $R(\phi)$ is finite then the map $\mathcal{R}(\phi) \to \mathcal{R}(\overline{\phi})$ is injective.*

PROOF OF LEMMA 31. By Mackey's Theorem (see [57]), every representation ρ of G is contained in a representation which is induced by a 1-dimensional representation χ of A. If ρ is fixed by $\hat{\phi}$ then for all $a \in A^0$ we have $\chi(a) = \chi(\phi(a))$. Let $A^1 = \{a \cdot \phi(a)^{-1} : a \in A^0\}$. By Lemma 30 $R(\phi|_{A^0})$ is finite and by Theorem 5 $R(\phi|_{A^0}) = [A^0 : A^1]$. Therefore A^1 has finite index in G. However we have shown that χ and therefore also ρ is constant on cosets of A^1. Therefore ρ has finite image, which implies that ρ is the restriction to G of a unique continuous irreducible representation $\overline{\rho}$ of $\overline{G}$. One verifies by continuity that $\hat{\overline{\phi}}(\overline{\rho}) = \overline{\rho}$.

Conversely if $\overline{\rho} \in S(\overline{\phi})$ then the restriction of $\overline{\rho}$ to G is in $S(\phi)$.

PROOF OF LEMMA 32. We must show that the intersection with G of the closure of $\{x\}_\phi$ in $\overline{G}$ is equal to $\{x\}_\phi$. We do this by constructing a coset of a normal subgroup of finite index in G which is contained in $\{x\}_\phi$. For every $a \in A^0$ we have $x \sim_\phi xa$ if there is a $b \in A^0$ with $x^{-1}bx\phi(b)^{-1} = a$. It follows that $\{x\}_\phi$ contains a coset of the group $A_x^2 := \{x^{-1}bx\phi(b)^{-1} : b \in A^0\}$. It remains to show that A_x^2 has finite index in G.

Let $\psi(g) = x\phi(g)x^{-1}$. Then by Corollary 4 we have $R(\psi) = R(\phi)$. This implies $R(\psi) < \infty$ and therefore by Lemma 30 that $R(\psi|_{A_0}) < \infty$. However by Theorem 5 we have $R(\psi|_{A_0}) = [A^0 : A_x^2]$. This finishes the proof.

Theorem 55 ([50]) *If ϕ be any endomorphism of G where G is an almost abelian group, then*

$$R(\phi) = \#\text{Fix}(\hat{\phi})$$

PROOF The proof follows from lemmas 31 , 32 and theorem 53

5.4 Endomorphisms of nilpotent groups

In this section, we shall extend the computation of the Reidemeister number to endomorphisms of finitely generated torsion free nilpotent groups via topological techniques. Let Γ be a finitely generated torsion free nilpotent group. It is well known [63] that $\Gamma = \pi_1(M)$ for some compact nilmanifold M. In fact, the *rank* (or *Hirsch number*) of Γ is equal to $dim M$, the dimension of M. Since M is a $K(\Gamma, 1)$, every endomorphism $\phi : \Gamma \to \Gamma$ can be realized by a selfmap $f : M \to M$ such that $f_\# = \phi$ and thus $R(f) = R(\phi)$.

Theorem 56 *Let Γ be a finitely generated torsion free nilpotent group of rank n. For any endomorphism $\phi : \Gamma \to \Gamma$ such that $R(\phi)$ is finite, there exists an endomorphism $\psi : \mathbb{Z}^n \to \mathbb{Z}^n$ such that $R(\phi) = \#\mathrm{Fix}\ \hat{\psi}$.*

PROOF: Let $f : M \to M$ be a map realizing ϕ on a compact nilmanifold M of dimension n. Following [21], M admits a principal torus bundle $T \to M \xrightarrow{p} N$ such that T is a torus and N is a nilmanifold of lower dimension. Since every selfmap of M is homotopic to a fibre preserving map of p, we may assume without loss of generality that f is fibre preserving such that the following diagram commutes.

$$
\begin{array}{ccc}
T & \xrightarrow{f_b} & T \\
\downarrow & & \downarrow \\
M & \xrightarrow{f} & M \\
p\downarrow & & \downarrow p \\
N & \xrightarrow{\bar{f}} & N
\end{array}
$$

A strengthened version of Anosov's theorem [3] is proven in [71] which states, in particular, that $|L(f)| = R(f)$ if $L(f) \neq 0$. Since the bundle p is orientable, the product formula $L(f) = L(f_b) \cdot L(\bar{f})$ holds and thus yields a product formula for the Reidemeister numbers, i.e., $R(f) = R(f_b) \cdot R(\bar{f})$. To prove the assertion, we proceed by induction on the rank of G or $dim M$.

The case where $n = 1$ follows from the theorem 52 since M is the unit circle . To prove the inductive step, we assume that $R(\bar{f}) = \#\mathrm{Fix}\ (\hat{\bar{\psi}} : \pi_1(\widehat{T^m})(= \widehat{\mathbb{Z}^m}) \to \pi_1(\widehat{T^m}))$ where $m < n$ and T^m is an m-torus. Since

$R(f_b) = \#\text{Fix} (\hat{\phi}_b : \widehat{\pi_1(T)} \to \widehat{\pi_1(T)})$, if $L(f) \neq 0$ then the product formula for Reidemeister numbers gives

$$
\begin{aligned}
R(f) &= \#\text{Fix} (\hat{\phi}_b) \cdot \#\text{Fix} (\widehat{\bar{\psi}}) \\
&= \#\text{Fix} (\hat{\psi})
\end{aligned}
$$

where $\psi = \phi_b \times \bar{\psi} : \pi_1(T^n) = \pi_1(T) \times \pi_1(T^m) \to \pi_1(T) \times \pi_1(T^m)$ with $T^n = T \times T^m$.

5.5   Main Theorem

The following lemma is useful for calculating Reidemeister numbers. It will also be used in the proof of the Main Theorem

Lemma 33 *Let $\phi : G \to G$ be any endomorphism of any group G, and let H be a subgroup of G with the properties*

$$\phi(H) \subset H$$

$$\forall x \in G \; \exists n \in I\!N \text{ such that } \phi^n(x) \in H.$$

Then

$$R(\phi) = R(\phi_H),$$

where $\phi_H : H \to H$ is the restriction of ϕ to H.

Proof Let $x \in G$. Then there is an n such that $\phi^n(x) \in H$. From Lemma 7 it is known that x is ϕ-conjugate to $\phi^n(x)$. This means that the ϕ-conjugacy class $\{x\}_\phi$ of x has non-empty intersection with H.

Now suppose that $x, y \in H$ are ϕ-conjugate, ie. there is a $g \in G$ such that

$$gx = y\phi(g).$$

We shall show that x and y are ϕ_H-conjugate, ie. we can find a $g \in H$ with the above property. First let n be large enough that $\phi^n(g) \in H$. Then applying ϕ^n to the above equation we obtain

$$\phi^n(g)\phi^n(x) = \phi^n(y)\phi^{n+1}(g).$$

This shows that $\phi^n(x)$ and $\phi^n(y)$ are ϕ_H-conjugate. On the other hand, one knows by Lemma 7 that x and $\phi^n(x)$ are ϕ_H-conjugate, and y and $\phi^n(y)$ are ϕ_H conjugate, so x and y must be ϕ_H-conjugate.

We have shown that the intersection with H of a ϕ-conjugacy class in G is a ϕ_H-conjugacy class in H. We therefore have a map

$$Rest: \begin{array}{ccc} \mathcal{R}(\phi) & \to & \mathcal{R}(\phi_H) \\ \{x\}_\phi & \mapsto & \{x\}_\phi \cap H \end{array}$$

This clearly has the two-sided inverse

$$\{x\}_{\phi_H} \mapsto \{x\}_\phi.$$

Therefore $Rest$ is a bijection and $R(\phi) = R(\phi_H)$.

Corollary 19 *Let $H = \phi^n(G)$. Then $R(\phi) = R(\phi_H)$.*

Let $\mu(d)$, $d \in I\!N$ be the Moebius function, i.e.

$$\mu(d) = \begin{cases} 1 & \text{if } d = 1, \\ (-1)^k & \text{if } d \text{ is a product of } k \text{ distinct primes}, \\ 0 & \text{if } d \text{ is not square} - \text{free}. \end{cases}$$

Theorem 57 (Congruences for the Reidemeister numbers) *Let $\phi : G \to G$ be an endomorphism of the group G such that all numbers $R(\phi^n)$ are finite and let H be a subgroup of G with the properties*

$$\phi(H) \subset H$$

$$\forall x \in G \; \exists n \in I\!N \text{ such that } \phi^n(x) \in H.$$

If one of the following conditions is satisfied:
(I) H is finitely generated Abelian,
(II) H is finite,
(III) H is a direct sum of a finite group and a finitely generated free Abelian group,
or more generally
(IV) H is finitely generated almost Abelian group,
or

(V) H is finitely generated, nilpotent and torsion free , then one has for all natural numbers n,

$$\sum_{d|n} \mu(d) \cdot R(\phi^{n/d}) \equiv 0 \bmod n.$$

PROOF From theorems 52, 55, 56 and lemma 33 it follows immediately that , in cases I - IV, for every n

$$R(\phi^n) = \#\text{Fix} \left[\hat{\phi}_H^{\ n} : \hat{H} \to \hat{H} \right].$$

Let P_n denote the number of periodic points of $\hat{\phi}_H$ of least period n. One sees immediately that

$$R(\phi^n) = \#\text{Fix} \left[\hat{\phi}_H^{\ n} \right] = \sum_{d|n} P_d.$$

Applying Möbius' inversion formula, we have,

$$P_n = \sum_{d|n} \mu(d) R(\phi^{n/d}).$$

On the other hand, we know that P_n is always divisible be n, because P_n is exactly n times the number of $\hat{\phi}_H$-orbits in $\hat{H}$ of length n. In the case V when H is finitely generated, nilpotent and torsion free ,we know from theorem 56 that there exists an endomorphism $\psi : \mathbf{Z}^n \to \mathbf{Z}^n$ such that $R(\phi^n) = \#\text{Fix}\ \hat{\psi}^n$. The proof then follows as in previous cases.

Remark 12 *For finite groups, congruences for Reidemeister numbers follow from those of Dold for Lefschetz numbers since we have identified in remark 2 the Reidemeister numbers with the Lefschetz numbers of induced dual maps.*

5.6   Congruences for Reidemeister numbers of a continuous map

Using corollary 1 we may apply the theorem 57 to the Reidemeister numbers of continuous maps.

Theorem 58 *Let* $f : X \rightarrow X$ *be a self-map such that all numbers* $R(f^n)$ *are finite. Let* $f_* : \pi_1(X) \rightarrow \pi_1(X)$ *be an induced endomorphism of the group* $\pi_1(X)$ *and let* H *be a subgroup of* $\pi_1(X)$ *with the properties*

$$f_*(H) \subset H$$

$$\forall x \in \pi_1(X) \ \exists n \in I\!N \ \text{such that} \ f_*^n(x) \in H.$$

If one of the following conditions is satisfied :
(I) H *is finitely generated Abelian,*
(II) H *is finite,*
(III) H *is a direct sum of a finite group and a finitely generated free Abelian group*
or more generally,
(IV) H *is finitely generated almost Abelian group,*
or
(V) H *is finitely generated, nilpotent and torsion free ,*
then one has for all natural numbers n,

$$\sum_{d|n} \mu(d) \cdot R(f^{n/d}) \equiv 0 \bmod n.$$

5.7 Congruences for Reidemeister numbers of equivariant group endomorphisms

Let G be a compact Abelian topological group acting on a topological group π as automorphisms of π, i.e., a homomorphism $\nu : G \rightarrow Aut(\pi)$. For every $\sigma \in \pi$, the isotropy subgroup of σ is given by $G_\sigma = \{g \in G | g(\sigma) = \sigma\}$ where $g(\sigma) = \nu(g)(\sigma)$. For any closed subgroup $H \leq G$, the fixed point set of the H-action, denoted by

$$\pi^H = \{\sigma \in \pi | h(\sigma) = \sigma, \forall h \in H\},$$

is a subgroup of π. Since G is Abelian, G acts on π^H as automorphisms of π^H. Denote by $\mathcal{C}(\pi)$ (and $\mathcal{C}(\pi^H)$) the set of conjugacy classes of elements of π (and π^H, respectively). Note that the group G acts on the conjugacy classes via

$$< \sigma > \mapsto < g(\sigma) >$$

for $< \sigma > \in \mathcal{C}(\pi)$ (or $\mathcal{C}(\pi^H)$).

Let $End_G(\pi)$ be the set of G-equivariant endomorphisms of π. For any $\phi \in End_G(\pi)$ and $H \leq G$, $\phi^H \in End_G(\pi^H)$ where $\phi^H = \phi|\pi^H : \pi^H \to \pi^H$. Furthermore, ϕ induces a G-map on $\mathcal{C}(\pi)$ defined by

$$\phi_{conj} : \mathcal{C}(\pi) \to \mathcal{C}(\pi)$$

via

$$< \sigma > \mapsto < \phi(\sigma) > .$$

Similarly, ϕ^H induces

$$(\phi^H)_{conj} : \mathcal{C}(\pi^H) \to \mathcal{C}(\pi^H).$$

In the case where π is Abelian, G acts on $\hat{\pi}$ via

$$\chi(\sigma) \mapsto \chi(g(\sigma))$$

for any $\chi \in \hat{\pi}, g \in G$. Thus the dual $\hat{\phi}$ of $\phi \in End_G(\pi)$ is also G-equivariant, i.e., $\hat{\phi} \in End_G(\hat{\pi})$. Similarly, $\widehat{\phi^H} \in End_G(\widehat{\pi^H})$.

To establish the congruence relations in this section, we need the following basic counting principle.

Lemma 34 *Let G be an Abelian topological group and Γ be a G-set with finite isotropy types. For any G-map $\psi : \Gamma \to \Gamma$,*

$$\text{Fix } \psi = \bigsqcup_{K \in Iso(\Gamma)} \text{Fix } \psi_K$$

where $Iso(\Gamma)$ is the set of isotropy types of Γ, $\Gamma_K = \{\gamma \in \Gamma | G_\gamma = K\}, \psi_K = \psi|\Gamma_K : \Gamma_K \to \Gamma^K$ and $\text{Fix } \psi_K = (\text{Fix } \psi) \cap \Gamma_K$. In particular, if $\#\text{Fix } \psi < \infty$ then

$$\#\text{Fix } \psi = \sum_{K \in Iso(\Gamma)} \#\text{Fix } \psi_K. \tag{5.2}$$

PROOF: It follows from the decomposition

$$\Gamma = \bigsqcup_{K \in Iso(\Gamma)} \Gamma_K. \square$$

Remark Suppose that G is a compact Abelian Lie group. If Fix ψ is finite then it follows that #Fix $\psi_K = I(\psi_K)$, the fixed point index of ψ_K which is divisible by $\chi(G/K)$, the Euler characteristic of G/K (see [16], [59], [96]).

Theorem 59 *Let G be an Abelian compact Lie group. For any subgroup $H \in Iso(\pi)$ and $\phi \in End_G(\pi)$, if*
  (I) *π is finitely generated and ϕ is eventually commutative or*
  (II) *π is finite or* (III) *π is finitely generated torsion free nilpotent, then*

$$\sum_{H \leq K \in Iso(\pi)} \mu(H, K) R(\phi^K) \equiv 0 \quad mod \ \chi(G/H)$$

and

$$\sum_{H \leq K \in Iso(\pi)} \varphi(H, K) R(\phi^K) \equiv 0 \quad mod \ \chi(G/H)$$

where $\mu(,)$ denotes the Möbius function on $Iso(\pi)$ and

$$\varphi(H, K) = \sum_{H \leq L \leq K} \chi(L/H)\mu(L, K).$$

PROOF: (I): Since ϕ is eventually commutative, so is ϕ^H for every $H \leq G$. With the canonical G-action on $H_1(\pi)$, $H_1(\phi)$ is a G-equivariant endomorphism of $H_1(\pi)$. Similarly, we have $H_1(\phi^H) \in End_G(H_1(\pi^H))$ and hence $\widehat{H_1(\phi^H)} \in End_G(\widehat{H_1(\pi^H)})$. It follows from theorem 52 and formula (5.1) that

$$R(\phi^H) = \#\text{Fix } (\widehat{H_1(\phi^H)}) = \sum_{H \leq K \in Iso(\pi)} \#\text{Fix } (\widehat{H_1(\phi^H)})_K.$$

Hence, by Möbius inversion [2], we have

$$\sum_{H \leq K \in Iso(\pi)} \mu(H, K) R(\phi^K) = \#\text{Fix } (\widehat{H_1(\phi^K)})_H \equiv 0 \quad mod \ \chi(G/H).$$

(II): Following theorem 14 , we have

$$R(\phi^H) = \#\text{Fix } (\phi^H)_{conj} = \sum_{H \leq K \in Iso(\pi)} \#\text{Fix } ((\phi^H)_{conj})_K.$$

The assertion follows from formula (5.1) and the Möbius inversion formula.

(III): The commutative diagram in the proof of Theorem 56 gives rise to the following commutative diagram in which the rows are short exact sequences of groups where π' is free abelian and $\bar{\pi}$ is nilpotent.

$$
\begin{array}{ccccccccc}
1 \to & \pi' & \xrightarrow{i_*} & \pi & \xrightarrow{p_*} & \bar{\pi} & \to 1 \\
 & \uparrow \phi' & & \uparrow \phi & & \uparrow \bar{\phi} \\
1 \to & \pi' & \xrightarrow{i_*} & \pi & \xrightarrow{p_*} & \bar{\pi} & \to 1
\end{array}
$$

Note that G acts on both π' and $\bar{\pi}$ as automorphisms and so ϕ' and $\bar{\phi}$ are both G-equivariant. Similarly, we have $\phi'^H \in End_G(\pi'^H)$ and $\bar{\phi}^H \in End_G(\bar{\phi}^H)$. It follows from the proof of Theorem 56 that the endomorphism ψ can be made G-equivariant by the diagonal action on the product of the fibre and the base. Hence,

$$R(\phi^H) = \#\text{Fix } \widehat{\psi^H} = \sum_{H \leq K \in Iso(\pi)} \#\text{Fix } (\widehat{\psi^H})_K.$$

Again, the assertion follows from Möbius inversion.

Finally, for the congruences with $\varphi(,)$, we proceed as in Theorem 6 of [59]. $\square$

Remark 13 *In [59], Komiya considered the G-invariant set $X^{(H)} = GX^H$ for arbitrary G (not necessarily Abelian). In our case, $\pi^{(H)}$ need not be a subgroup of π. For example, take $\pi = G = S_3$ to be the symmetric group on three letters and let G act on π via conjugation. For any subgroup $H \leq G$ of order two, it is easy to see that $\pi^{(H)}$ is not a subgroup of π.*

5.8 Congruences for Reidemeister numbers of equivariant maps

Unlike Komiya's generalization [59] of Dold's result, the congruence relations among the Reidemeister numbers of f^n established in section cannot be generalized without further assumptions as we illustrate in the following example.

Example 9 Let $X = S^2 \subset \mathbf{R}^3, G = \mathbf{Z}_2$ act on S^2 via

$$\zeta(x_1, x_2, x_3) = (x_1, x_2, \zeta x_3)$$

so that $X^G = S^1$. Let $f : X \to X$ be defined by $(x_1, x_2, x_3) \mapsto (-x_1, x_2, x_3)$. It follows that

$$R(f) = 1; \qquad R(f^G) = 2.$$

Hence,

$$\sum_{(1) \leq K} \mu((1), K) R(f^K) = \mu(1) R(f) + \mu(2) R(f^G) = -1$$

which is not congruent to 0 *mod* 2.

In order to apply the previous result , we need $R(f^H) = R((f_*)^H)$, for all $H \leq G$. In the above example, $R(f^G) = 2$ but $R((f_*)^G) = 1$.

Recall that a selfmap $f : X \to X$ is *eventually commutative* [51] if the induced homomorphism $f_* : \pi_1(X) \to \pi_1(X)$ is eventually commutative. The following is immediate from Theorem 59.

Theorem 60 *Let* $f : X \to X$ *be a* G-*map on a finite* G-*complex* X *where* G *is an abelian compact Lie group, such that* X^H *is connected for all* $H \in Iso(X)$. *If* f *is eventually commutative, or* $\pi_1(X^H)$ *is finite or nilpotent, and* $R(f^H) = R((f_*)^H)$ *for all* $H \in Iso(X)$, *then*

$$\sum_{H \leq K \in Iso(\pi)} \mu(H, K) R(f^K) \equiv 0 \quad mod \ \chi(G/H)$$

and

$$\sum_{H \leq K \in Iso(\pi)} \varphi(H, K) R(f^K) \equiv 0 \quad mod \ \chi(G/H).$$

5.9   Congruences for Nielsen numbers of a continuous map

Theorem 61 *Suppose that there is a natural number m such that $\tilde{f}_*^m(\pi) \subset I(\tilde{f}^m)$. If for every d dividing a certain natural number n we have $L(f^{n/d}) \neq 0$, then one has for that particular n,*

$$\sum_{d|n} \mu(d) N(f^{n/d}) \equiv 0 \bmod n.$$

PROOF From the results of Jiang [51] we have that $N(f^{n/d}) = R(f^{n/d})$ for the same particular n and $\tilde{f}_*$ is eventually commutative. The result now follows from theorem 58.

Corollary 20 *Let $I(id_{\tilde{X}}) = \pi$ and for every d dividing a certain natural number n we have $L(f^{n/d}) \neq 0$, then theorem 61 applies*

Corollary 21 *Suppose that X is aspherical, f is eventually commutative and for every d dividing a certain natural number n we have $L(f^{n/d}) \neq 0$, then theorem 61 applies*

Example 10 *Let $f : T^n \to T^n$ be a hyperbolic endomorphism. Then for every natural n*

$$\sum_{d|n} \mu(d) N(f^{n/d}) \equiv 0 \bmod n.$$

Example 11 *Let $g : M \to M$ be an expanding map [86] of the orientable smooth compact manifold M. Then M is aspherical and is a $K(\pi_1(M), 1)$, and $\pi_1(M)$ is torsion free [86]. According to Shub [86] any lifting $\tilde{g}$ of g has exactly one fixed point. From this and the covering homotopy theorem it follows that the fixed point of g are pairwise inequivalent. The same is true for all iterates g^n. Therefore $N(g^n) = \#\mathrm{Fix}\,(g^n)$ for all n. So the sequence of the Nielsen numbers $N(g^n)$ of an expanding map satisfies the congruences as above.*

Theorem 62 *Let X be a connected, compact polyhedron with finite fundamental group π. Suppose that the action of π on the rational homology of the universal cover $\tilde{X}$ is trival, i.e. for every covering translation $\alpha \in \pi$,*

$\alpha_* = id : H_*(\tilde{X}, \mathbb{Q}) \rightarrow H_*(\tilde{X}, \mathbb{Q})$. If for every d dividing a certain natural number n we have $L(f^{n/d}) \neq 0$, then one has for that particular n,

$$\sum_{d|n} \mu(d) N(f^{n/d}) \equiv 0 \bmod n.$$

PROOF From the results of Jiang [51] we have that $N(f^{n/d}) = R(f^{n/d})$ for the same particular n. The result now follows from theorem 58.

Lemma 35 *Let X be a polyhedron with finite fundamental group π and let $p : \tilde{X} \rightarrow X$ be its universal covering. Then the action of π on the rational homology of $\tilde{X}$ is trivial iff $H_*(\tilde{X}; \mathbb{Q}) \cong H_*(X; \mathbb{Q})$.*

Corollary 22 *Let $\tilde{X}$ be a compact 1-connected polyhedron which is a rational homology n-sphere, where n is odd. Let π be a finite group acting freely on $\tilde{X}$ and let $X = \tilde{X}/\pi$. Then theorem 62 applies.*

PROOF The projection $p : \tilde{X} \rightarrow X = \tilde{X}/\pi$ is a universal covering space of X. For every $\alpha \in \pi$, the degree of $\alpha : \tilde{X} \rightarrow \tilde{X}$ must be 1, because $L(\alpha) = 0$ (α has no fixed points). Hence $\alpha_* = id : H_*(\tilde{X}; \mathbb{Q}) \rightarrow H_*(\tilde{X}; \mathbb{Q})$.

Corollary 23 *If X is a closed 3-manifold with finite π, then theorem 62 applies.*

PROOF $\tilde{X}$ is an orientable, simply connected manifold, hence a homology 3-sphere. We apply corollary 22.

Example 12 *Let $X = L(m, q_1, \ldots, q_r)$ be a generalized lens space and $f : X \rightarrow X$ a continuous map with $f_{1*}(1) = k$ where $|k| \neq 1$. Then for every natural n*

$$\sum_{d|n} \mu(d) N(f^{n/d}) \equiv 0 \bmod n.$$

PROOF

By corollary we see that theorem applies for lens spaces. Since $\pi_1(X) = \mathbb{Z}/m\mathbb{Z}$, the map f is eventually commutative. A lens space has a structure as a CW complex with one cell e_i in each dimension $0 \leq i \leq 2n + 1$. The boundary map is given by $\partial e_{2k} = m.e_{2k-1}$ for even cells, and $\partial e_{2k+1} = 0$ for odd cells. From this we may calculate the Lefschetz numbers:

$$L(f^n) = 1 - k^{(l+1)n} \neq 0.$$

This is true for any n as long as $| k | \neq 1$. The result now follows from theorem .

Remark 14 *It is known that in previous example*

$$N(f^n) = R(f^n) = \#\text{Coker}\,(1 - f_{1*}^n) = hcf(1 - k^n, m)$$

for every n.So we obtain pure arithmetical fact: the sequence $n \mapsto hcf(1 - k^n, m)$ satisfies congruences above for every natural n if $| k | \neq 1$.

5.10 Some conjectures for wider classes of groups

For the case of almost nilpotent groups (ie. groups with polynomial growth, in view of Gromov's theorem [44]) we believe that the congruences for the Reidemeister numbers are also true.We intend to prove this conjecture by identifying the Reidemeister number on the nilpotent part of the group with the number of fixed points in the direct sum of the duals of the quotients of successive terms in the central series. We then hope to show that the Reidemeister number of the whole endomorphism is a sum of numbers of orbits of such fixed points under the action of the finite quotient group (ie the quotient of the whole group by the nilpotent part). The situation for groups with exponential growth is very different. There one can expect the Reidemeister number to be infinite as long as the endomorphism is injective.

Chapter 6

The Reidemeister torsion

6.1 Preliminaries

Like the Euler characteristic, the Reidemeister torsion is algebraically defined. Roughly speaking, the Euler characteristic is a graded version of the dimension, extending the dimension from a single vector space to a complex of vector spaces. In a similar way, the Reidemeister torsion is a graded version of the absolute value of the determinant of an isomorphism of vector spaces: for this to make sense, both vector spaces should be equipped with a positive density.

Recall that a density f on a complex space V of dimension n is a map $f : \wedge^n V \to R$ with $f(\lambda \cdot x) = |\lambda| \cdot f(x)$ for all $x \in \wedge^n V, \lambda \in \mathcal{C}$. The densities on V clearly form a real vector space $|V|$ of dimension one. If f is nonzero and takes values in $[o, \infty)$, we say that it is positive . If V_1 and V_2 are both n-dimensional and $A : V_1 \to V_2$ is linear over $\mathcal{C}$, then there is an induced map $A^* : |V_2| \to |V_1|$. If each V_i carries a preffered positive density f_i then $A^* f_2 = a \cdot f_1$, for some $a \geq 0$, and we write $a = |det A|$.In the case $V_1 = V_2, f_1 = f_2$ then a is indeed the absolute value of the determinant of A, so this notation is consistent.

If $0 \to V_1 \to V_2 \to V_3 \to 0$ is an exact sequence of finite dimensional vector spaces over $\mathcal{C}$, then there is a natural isomorphism $\wedge^{n_1} V_1 \otimes \wedge^{n_3} V_3 \to \wedge^{n_2} V_2$, $n_i = dim V_i$.This induces a natural isomorphism $|V_1| \otimes |V_3| \cong |V_2|$. If V is zero, then absolute value is a standard generator for $|V|$ and sets up a natural isomorphism $|V| \cong R$.

113

Let $d^i : C^i \to C^{i+1}$ be a cochain complex of finite-dimensional vector spaces over $\mathbb{C}$ with $C^i = 0$ for $i < 0$ or i large. Let $Z^i = \ker d^i$, $B^i = \operatorname{im} d^{i-1}$, and $H^i = Z^i / B^i$ be the cocycles, coboundaries, and cohomology of C^* respectively. If one is given positive densities Δ_i on C^i and D_i on H^i(with the standard choice when i is negative or large), then the Reidemeister torsion $\tau(C^*, \Delta_i, D_i) \in (0, \infty)$ is defined as follows. The short exact sequences $0 \to Z^i \to C^i \to B^{i+1} \to 0$ and $0 \to B^i \to Z^i \to H^i \to o$ together with the trivializations $|C^i| = R \cdot \Delta_i, |H^i| = R \cdot D_i$ give isomorphisms $|Z^i| \otimes |B^{i+1}| \cong R$, $|B^i| \cong |Z^i|$ for each i. This gives isomorphisms $|B^i| \otimes |B^{i+1}| \cong R$, hence also $|B^i| \cong |B^{i+2}|$. But for $i \leq 0$ or i large, $B^i = 0$ and so $|B^i| = R$. So if j is large and even, we find $R = |B^j| \cong |B^0| = R$ and this isomorphism is the multiplication by scalar $\tau \in R^*$. This is the Reidemeister torsion $\tau(C^*, \Delta_i, D_i)$.

If the cohomology $H^i = 0$ for all i we say that C^* is *acyclic* and we write $\tau(C^*, \Delta_i)$ for the torsion when all the D_i are standard .

When C^* is acyclic , there is another way to define τ.

Definition 15 *Consider a chain contraction $\delta^i : C^i \to C^{i-1}$, i.e. a linear map such that $d \circ \delta + \delta \circ d = id$. Then $d + \delta$ determies a map*

$(d + \delta)_+ : C^+ := \oplus C^{2i} \to C^- := \oplus C^{2i+1}$ *and a map* $(d + \delta)_- : C^- \to C^+$. *Since the map $(d+\delta)^2 = id + \delta^2$ is unipotent, $(d+\delta)_+$ must be an isomorphism. One defines $\tau(C^*, \Delta_i) := | \det(d + \delta)_+ |$ (see [42]).*

Reidemeister torsion is defined in the following geometric setting. Suppose K is a finite complex and E is a flat, finite dimensional, complex vector bundle with base K. We recall that a flat vector bundle over K is essentially the same thing as a representation of $\pi_1(K)$ when K is connected. If $p \in K$ is a basepoint then one may move the fibre at p in a locally constant way around a loop in K. This defines an action of $\pi_1(K)$ on the fibre E_p of E above p. We call this action the holonomy representation $\rho : \pi \to GL(E_p)$. Conversely, given a representation $\rho : \pi \to GL(V)$ of π on a finite dimensional complex vector space V, one may define a bundle $E = E_\rho = (\tilde{K} \times V)/\pi$. Here $\tilde{K}$ is the universal cover of K, and π acts on $\tilde{K}$ by covering tranformations and on V by ρ. The holonomy of E_ρ is ρ, so the two constructions give an equivalence of flat bundles and representations of π.

If K is not connected then it is simpler to work with flat bundles. One then defines the holonomy as a representation of the direct sum of π_1 of the components of K. In this way, the equivalence of flat bundles and representations is recovered.

Suppose now that one has on each fibre of E a positive density which is locally constant on K. In terms of ρ_E this assumption just means $|\det \rho_E| = 1$. Let V denote the fibre of E. Then the cochain complex $C^i(K; E)$ with coefficients in E can be identified with the direct sum of copies of V associated to each i-cell σ of K. The identification is achieved by choosing a basepoint in each component of K and a basepoint from each i-cell. By choosing a flat density on E we obtain a preferred density Δ_i on $C^i(K, E)$. Let $H^i(K; E)$ be the i-dimensional cohomology of K with coefficients in E , i.e. the twisted cohomology of E. Given a density D_i on each $H^i(K; E)$, one defines the Reidemeister torsion of $(K; E, D_i)$ to be $\tau(K; E, D_i) = \tau(C^*(K; E), \Delta_i, D_i) \in (0, \infty)$. A case of particular interest is when E is an acyclic bundle, meaning that the twisted cohomology of E is zero $(H^i(K; E) = 0)$, then one can take D_i to be the absolute value map on $\wedge^0(0) = \mathbb{C}$ and the resulting Reidemeister torsion is denoted by $\tau(K; E)$. In this case it does not depend on the choice of flat density on E.

The Reidemeister torsion of an acyclic bundle E on K has many nice properties. Suppose that A and B are subcomplexes of K. Then we have a multiplicative law:

$$\tau(A \cup B; E) \cdot \tau(A \cap B; E) = \tau(A; E) \cdot \tau(B; E) \qquad (6.1)$$

that is interpreted as follows. If three of the bundles $E|A \cup B$, $E|A \cap B$, $E|A$, $E|B$ are acyclic then so is the fourth and the equation (6.1) holds.

Another property is the simple homotopy invariance of the Reidemeister torsion. Suppose K' is a subcomplex of K obtained by an elementary collapse of an n-cell σ in K. This means that $K = K' \cup \sigma \cup \sigma'$ where σ' is an $(n-1)$ cell of K so set up that $\partial \sigma' = \sigma' \cap K'$ and $\sigma' \subset \partial \sigma$, i.e. σ' is a free face of σ. So one can push σ' through σ into K' giving a homotopy equivalence. Then $H^*(K; E) = H^*(K'; E)$ and

$$\tau(K; E, D_i) = \tau(K'; E, D_i) \qquad (6.2)$$

By iterating a sequence of elementary collapses and their inverses, one obtains a homotopy equivalence of complexes that is called *simple* . Plainly one has, by iterating (6.2) , that the Reidemeister torsion is a simply homotopy invariant. In particular τ is invariant under subdivision. This implies that for a smooth manifold, one can unambiguously define $\tau(K; E, D_i)$ to be the torsion of any smooth triangulation of K.

In the case $K = S^1$ is a circle, let A be the holonomy of a generator of the fundamental group $\pi_1(S^1)$. One has that E is acyclic iff $I - A$ is invertible and then

$$\tau(S^1; E) = |\det(I - A)| \tag{6.3}$$

Note that the choice of generator is irrelevant as $I - A^{-1} = (-A^{-1})(I - A)$ and $|det(-A^{-1})| = 1$.

These three properties of the Reidemeister torsion are the analogues of the properties of Euler characteristic (cardinality law, homotopy invariance and normalization on a point), but there are differences.Since a point has no acyclic representations ($H^0 \neq 0$) one cannot normalise τ on a point as we do for the Euler characteristic, and so one must use S^1 instead. The multiplicative cardinality law for the Reidemeister torsion can be made additive just by using $\log \tau$, so the difference here is inessential. More important for some purposes is that the Reidemeister torsion is not an invariant under a general homotopy equivalence: as mentioned earlier this is in fact why it was first invented.

It might be expected that the Reidemeister torsion counts something geometric(like the Euler characteristic). D. Fried showed that it counts the periodic orbits of a flow and the periodic points of a map. We will show that the Reidemeister torsion counts the periodic point classes of a map(fixed point classes of the iterations of the map).

Some further properties of τ describe its behavior under bundles.

Let $p : X \to B$ be a simplicial bundle with fiber F where F, B, X are finite complexes and p^{-1} sends subcomplexes of B to subcomplexes of X . over the circle S^1. We assume here that E is a flat, complex vector bundle over B . We form its pullback p^*E over X. Note that the vector spaces $H^i(p^{-1}(b), \mathbb{C})$ with $b \in B$ form a flat vector bundle over B, which we denote H^iF. The integral lattice in $H^i(p^{-1}(b), \mathbb{R})$ determines a flat density by the condition that the covolume of the lattice is 1. We suppose that the bundle $E \otimes H^iF$ is acyclic for all i. Under these conditions D. Fried [42] has shown that the bundle p^*E is acyclic, and

$$\tau(X; p^*E) = \prod_i \tau(B; E \otimes H^iF)^{(-1)^i}. \tag{6.4}$$

the opposite extreme is when one has a bundle E on X for which the restriction $E|F$ is acyclic. Then, for B connected,

$$\tau(X; E) = \tau(F; E|F)^{\chi(B)} \tag{6.5}$$

Suppose in (6.5) that $F = S^1$ i.e. X is a circle bundle . Then (6.5) can be regarded as saing that

$$\log \tau(X; E) = \chi(B) \cdot \log \tau(F; E|F)$$

is counting the circle fibers in X in the way that χ counts points in B, with a weighting factor of $\log \tau(F; E|F)$.

6.2   The Reidemeister zeta Function and the Reidemeister Torsion of the Mapping Torus of the dual map.

Let $f : X \to X$ be a homeomorphism of a compact polyhedron X. Let $T_f := (X \times I)/(x, 0) \sim (f(x), 1)$ be the mapping tori of f. We shall consider the bundle $p : T_f \to S^1$ over the circle S^1. We assume here that E is a flat, complex vector bundle with finite dimensional fibre and base S^1. We form its pullback p^*E over T_f. Note that the vector spaces $H^i(p^{-1}(b), c)$ with $b \in S^1$ form a flat vector bundle over S^1, which we denote H^iF. The integral lattice in $H^i(p^{-1}(b), \mathbb{R})$ determines a flat density by the condition that the covolume of the lattice is 1. We suppose that the bundle $E \otimes H^iF$ is acyclic for all i. Under these conditions D. Fried [42] has shown that the bundle p^*E is acyclic, and we have

$$\tau(T_f; p^*E) = \prod_i \tau(S^1; E \otimes H^iF)^{(-1)^i}. \tag{6.6}$$

Let g be the prefered generator of the group $\pi_1(S^1)$ and let $A = \rho(g)$ where $\rho : \pi_1(S^1) \to GL(V)$. Then the holonomy around g of the bundle $E \otimes H^iF$ is $A \otimes f_i^*$.

Since $\tau(S^1; E) = |\det(I - A)|$ it follows from (6.6) that

$$\tau(T_f; p^*E) = \prod_i |\det(I - A \otimes f_i^*)|^{(-1)^i}. \tag{6.7}$$

We now consider the special case in which E is one-dimensional, so A is just a complex scalar λ of modulus one. Then in terms of the rational function $L_f(z)$ we have [42]:

$$\tau(T_f; p^*E) = \prod_i \mid \det(I - \lambda \cdot f_i^*) \mid^{(-1)^i} = \mid L_f(\lambda) \mid^{-1} \qquad (6.8)$$

Theorem 63 *Let $\phi : G \to G$ be an automorphism of G, where G is the direct sum of a finite group with a finitely generated free Abelian group, then*

$$\tau\left(T_{\hat{\phi}}; p^*E\right) = \mid L_{\hat{\phi}}(\lambda) \mid^{-1} = \mid R_\phi(\sigma \cdot \lambda) \mid^{(-1)^{r+1}}, \qquad (6.9)$$

where λ is the holonomy of the one-dimensional flat complex bundle E over S^1, r and σ are the constants described in theorem 13 .

PROOF We know from the theorem 52 that $R(\phi^n)$ is the number of fixed points of the map $\hat{\phi}^n$. In general it is only necessary to check that the number of fixed points of $\hat{\phi}^n$ is equal to the absolute value of its Lefschetz number. We assume without loss of generality that $n = 1$. We are assuming that $R(\phi)$ is finite, so the fixed points of $\hat{\phi}$ form a discrete set. We therefore have

$$L(\hat{\phi}) = \sum_{x \in \text{Fix } \hat{\phi}} \text{Index } (\hat{\phi}, x).$$

Since ϕ is a group endomorphism, the trivial representation $x_0 \in \hat{G}$ is always fixed. Let x be any fixed point of $\hat{\phi}$. Since $\hat{G}$ is union of tori $\hat{G}_0, ..., \hat{G}_t$ and $\hat{\phi}$ is a linear map, we can shift any two fixed points onto one another without altering the map $\hat{\phi}$. This gives us for any fixed point x the equality

$$\text{Index } (\hat{\phi}, x) = \text{Index } (\hat{\phi}, x_0)$$

and so all fixed points have the same index. It is now sufficient to show that Index $(\hat{\phi}, x_0) = \pm 1$. This follows because the map on the torus

$$\hat{\phi} : \hat{G}_0 \to \hat{G}_0$$

lifts to a linear map of the universal cover, which is an euclidean space. The index is then the sign of the determinant of the identity map minus this lifted map. This determinant cannot be zero, because $1 - \hat{\phi}$ must have finite kernel by our assumption that the Reidemeister number of ϕ is finite (if $\det(1 - \hat{\phi}) = 0$ then the kernel of $1 - \hat{\phi}$ is a positive dimensional subspace of $\hat{G}$, and therefore infinite).

Corollary 24 *Let $f : X \to X$ be a homeomorphism of a compact polyhedron X. If $\pi_1(X)$ is the direct sum of a finite group with a free Abelian group, then then*

$$\tau\left(T_{\widehat{(f_{1*})}}; p^*E\right) = \left| L_{\widehat{(f_{1*})}}(\lambda) \right|^{-1} = \left| R_f(\sigma \cdot \lambda) \right|^{(-1)^{r+1}},$$

where r and σ are the constants described in theorem 13 .

6.3 The connection between the Reidemeister torsion, eta–invariant, the Rochlin invariant and theta multipliers via the dynamical zeta functions

In this section we establish a connection between the Reidemeister torsion of a mapping tori, the eta-invariant, the Rochlin invariant, and theta multipliers via the Lefchetz zeta function and the Bismut-Freed-Witten holonomy theorem.

6.3.1 Rochlin invariant

We begin by recalling the definition of a spin structure on an oriented Riemannian manifold M^m with special attention to the notion of a spin diffeomorphism. The tangent bundle TM is associated to a principal $GL_+(m)$ bundle P, the bundle of oriented tangent frames. This last group $GL_+(m)$ has a unique connected two-fold covering group $\widetilde{GL}_+(m)$. If the tangent bundle TM is associated to a princial $\widetilde{GL}_+(m)$ bundle $\tilde{P}$, then we call $\tilde{P}$ a *spin structure* on M. Thus, spin structures are in one-to-one correspondence with double coverings $\tilde{P} \to P$ which are nontrivial on each fiber. In terms of cohomology, such coverings are given by elements w of $H^1(P, \mathbb{Z}/2\mathbb{Z})$ such that $w \mid \pi^{-1}(x) = 0$ for every x in M, where $\pi : P \to M$ is the bundle projection and $\pi^{-1}(x)$ is the fiber over x. We shall refer to the cohomology class w as a spin structure on M, and the pair (M, w) as a spin manifold.

Given an orientation preserving diffeomorphism $f : M \to M$, the differential df of f gives us a diffeomorphism $df : P \to P$ and hence an isomorphism on the cohomology $(df)^* : H^1(P, \mathbb{Z}/2\mathbb{Z}) \to H^1(P, \mathbb{Z}/2\mathbb{Z})$. We say that an

orientation-preserving diffeomorphism $f : M \to M$ preserves the spin structure w if $(df)^*(w) = w$ in $H^1(P, \mathbb{Z}/2\mathbb{Z})$. This is equivalent to the existence
of a bundle map $b : \tilde{P} \to \tilde{P}$ making the following diagram commutative

$$
\begin{array}{ccc}
\tilde{P} & \xrightarrow{b} & \tilde{P} \\
\downarrow & & \downarrow \\
P & \xrightarrow{df} & P
\end{array}
$$

By a spin diffeomorphism F of (M, w) we mean a pair $F = (f, b)$ consisting
not only of a spin preserving diffeomorphism f but also of a bundle map
$b : \tilde{P} \to \tilde{P}$ covering df. Given a spin diffeomorphism $F = (f, b)$ of (M, w)
there is a well defined spin structure w' on the mapping torus T_f (see [61]).

We shall now define an invariant of spin diffeomorphisms F of (M, w)
where M has dimension $8k + 2$. This will actually be defined via (T_f, w'),
a spin manifold of dimension $8k + 3$. As usual, a spin manifold (N^{8k+3}, w)
is a spin boundary if there is a compact $8k + 4$-dimensional spin manifold
(X^{8k+4}, W) with $\partial X = N$ and with W restricting to w. By a result of [3]
the manifold N^{8k+3} is a spin boundary if and only if it is an unoriented
boundary. A necessary and sufficient condition for N to be a boundary is
that all its Stiefel-Whitney numbers vanish. In particular, this is always
the case when $k = 0$ or $k = 1$. If M^{8k+2} has vanishing Stiefel-Whitney
classes, then all the Stiefel-Whitney numbers of T_f are zero (see [61]). For a
spin boundary $(N^{8k+3}, w) = \partial(X^{8k+4}, W)$ the *Rochlin invariant* $R(N, w)$ in
$\mathbb{Z}/16\mathbb{Z}$ is defined by

$$
R(N, w) \equiv \sigma(X) \pmod{16},
$$

where $\sigma(X)$ denotes the signature of the $8k + 4$-dimensional manifold X.

6.3.2 Determinant line bundles, the Eta-invariant and the Bismut-Freed-Witten theorem

Let $\pi : Z \to N$ be a smooth fibration of manifolds with base manifold N and
total manifold Z. The fiber above a point $x \in N$ is an $8k + 2$-dimensional
manifold M_x^{8k+2} which is equipped with a metric and a compatible spin structure. The latter vary smoothly with respect to the parameter x in the base
manifold; in other words the structure group of the fibration $\pi : Z \to N$ is a
subgroup of the spin diffeomorphism group.

In this situation, along a fiber M_x, we have a principal $Spin(8k + 2)$-bundle $P(TM_x)$. Since the dimension $8k + 2$ is even there are two half-spin representation $S_\pm$ of $Spin(8k + 2)$ and associated to them two vector bundles $E_x^\pm = P(TM_x) \otimes S^\pm$. On the space $C^\infty(E_x^\pm)$ of C^∞-sections of these bundles, there is a Dirac operator $\partial_x : C^\infty(E_x^\pm) \to C^\infty(E_x^\mp)$ which is a first order, elliptic, differential operator [18]. If we replace the C^∞-sections of $E_x^\pm$ by square-integrable sections, the Dirac operator can be extended to an operator $\partial_x : L^2(E_x^\pm) \to L^2(E_x^\mp)$ of Hilbert spaces. As we vary x over N, these Hilbert spaces $L^2(E_x^\pm)$ form Hilbert bundles $L^2(E^\pm)$ and the operators ∂_x form a continuous family of operators $\partial : L^2(E^\pm) \to L^2(E^\mp)$ on these Hilbert bundles.

From the work Atiyah-Singer, Bismut-Freed [9] and Quillen [76] it follows that there exists a well-defined complex line bundle $\det \partial$ over N. Over a point x in N, the fiber of this line bundle $(\det \partial)_x$ is isomorphic to $(\Lambda^{max} ker\partial_x)^* \otimes (\Lambda^{max} coker\partial_x)$. Bismut and Freed [9] undertook an extensive study of the geometry of this determinant line bundle $\det \partial$. They showed that $\det \partial$ admits a Bismut-Freed connection $\triangledown$, and proved a formula for the curvature associated to this connection. One of the basic results in [9] is the holonomy formula for the Bismut-Freed connection $\triangledown$ of the determinant line bundle $\det \partial$ around an immersed circle $\gamma : S^1 \to N$ in the base manifold N. We now describe this formula. Pulling back by γ, there is an $8k + 3$-dimensional manifold which is diffeomorphic to a mapping torus T_ϕ, with the diffeomorphism ϕ specified by γ. Choosing an arbitrary metric g_{S^1} on S^1, and using the projection $\Phi : \tau(T_\phi) \to \tau_{fiber}(T_\phi)$ of tangent bundles, we obtain a Riemannian structure on T_ϕ. Since the structure group of the fibration $\pi : Z \to N$ is a subgroup of the spin diffeomorphism group, it follows that ϕ is covered by a canonical spin diffeomorphism and the mapping torus T_ϕ has a natural spin structure. From this spin structure on T_ϕ we obtain a spin bundle over T_ϕ with structure group $Spin(8k + 3)$ and Dirac operator ∂ on the space of C^∞-sections of this bundle.

Following Atiyah, Patodi, Singer [7] we define the function $\eta(s, \partial)$ in terms of the eigenvalues λ of ∂ by

$$\eta(s, \partial) = \sum_{\lambda \neq 0} \frac{sign\lambda}{|\lambda|^s}.$$

This function is holomorphic for $Re(s) > 0$ and its value at $s = 0$ is the η-invariant of ∂: $\eta(\partial) = \eta(0, \partial)$. We denote by $h(\partial)$ the dimension of the kernel of the operator ∂. Notice that these invariants depend on the choice of the metric g_{S^1} on the base circle S^1. In order to be free of this choice, we scale the metric g_{S^1} by a factor $\frac{1}{\epsilon^2}$, and with respect to this new metric $\frac{g_{S^1}}{\epsilon^2}$ we have a Dirac operator ∂_ϵ on T_ϕ and a corresponding η-invariant $\eta(\partial_\epsilon)$ and $h(\partial_\epsilon)$. As the parameter ϵ tends to zero, the invariant $\frac{\eta(\partial_\epsilon)+h(\partial_\epsilon)}{2}$ tends to a fixed limit.

Theorem 64 *[9] The holonomy of the Bismut-Freed connection $\bigtriangledown$ of the determinant line bundle $\det \partial$ around γ is given by*

$$hol(\gamma; \det \partial, \bigtriangledown) = \lim_{\epsilon \to 0} \exp\left(-2\pi i \frac{\eta(\partial_\epsilon) + h(\partial_\epsilon)}{2}\right). \qquad (6.10)$$

Now, suppose that $\pi : Z \to N$ is a fibration of manifolds M^{8k+2} with two prefered spin structures w_1 and w_2. Corresponding to these two spin structures, there are families of Dirac operators ∂_{w_1} and ∂_{w_2}, and determinant line bundles $\det \partial_{w_1}$ and $\det \partial_{w_2}$. Notice that from the curvature formula of Bismut-Freed [9] these two complex line bundles $\det \partial_{w_1}$ and $\det \partial_{w_2}$ have the same curvature 2-form. Hence if we form the bundle $\det \partial_{w_1} / \det \partial_{w_2} = \det \partial_{w_1} \otimes (\det \partial_{w_2})^*$, the result is a flat complex line bundle. In the language of contemporary physicists, this is known as the cancelation of local anomalies. For some of their models, it is important to investigate the holonomies of the flat line bundle $\det \partial_{w_1} \otimes (\det \partial_{w_2})^*$ -the global anomalies.

Let w_1' and w_2' be the spin structures on T_ϕ induced by w_1 and w_2, where the diffeomorphism ϕ is specified by γ. Lee, Miller and Weintraub proved the following

Theorem 65 ([61]) *The holonomy of the flat complex line bundle* $\det \partial_{w_1} / \det \partial_{w_2}$ *around a loop γ is*

$$hol(\gamma; \det \partial_{w_1} / \det \partial_{w_2}) =$$

$$= \lim_{\epsilon \to 0} \exp\left(-2\pi i \left[\frac{\eta(\partial_{w_1'\epsilon}) + h(\partial_{w_1'\epsilon})}{2} - \frac{\eta(\partial_{w_2'\epsilon}) + h(\partial_{w_2'\epsilon})}{2}\right]\right)$$

If the fiber M^{8k+2} is a Riemann surface, then

$$hol(\gamma; \det \partial_{w_1} / \det \partial_{w_2}) = \exp\left[-2\pi i \frac{R(T_\phi, w_1') - R(T_\phi, w_2')}{8}\right].$$

6.3.3   Connection with Reidemeister torsion

Let us now suppose that $F = (f, b_i)$, $i = 1, 2$ are two spin diffeomorphisms of spin manifolds (M^{8k+2}, w_i), $i = 1, 2$. Then there are well-defined spin structures w_i', $i = 1, 2$ on the mapping torus T_f. We may consider T_f as a bundle $p : T_f \to S^1$ over the circle S^1 with fiber M^{8k+2}. As above we have two Dirac operators $\partial_{w_i' \epsilon}$, $i = 1, 2$ on the space of C^∞-sections of the spin bundles over T_f and corresponding eta-invariants $\eta(\partial_{w_i' \epsilon})$ and $h(\partial_{w_i' \epsilon})$. We also have a flat complex determinant line bundle $\det \partial_{w_1} / \det \partial_{w_2}$ over the base manifold S^1 and its pullback $p^*(\det \partial_{w_1} / \det \partial_{w_2})$ over T_f. Let γ be the preferred generator of the fundamental group $\pi_1(S^1)$.

Theorem 66

$$\tau(T_f; p^*(\det \partial_{w_1} / \det \partial_{w_2})) = \mid L_f(\lambda) \mid^{-1},$$

where

$$\lambda = hol(\gamma; \det \partial_{w_1} / \det \partial_{w_2}) =$$

$$= \lim_{\epsilon \to 0} \exp\left(-2\pi i \left[\frac{\eta(\partial_{w_1' \epsilon}) + h(\partial_{w_1' \epsilon})}{2} - \frac{\eta(\partial_{w_2' \epsilon}) + h(\partial_{w_2' \epsilon})}{2}\right]\right)$$

If the fiber M^{8k+2} is a Riemann surface, then

$$\lambda = hol(\gamma; \det \partial_{w_1} / \det \partial_{w_2}) = \exp\left[-2\pi i \frac{R(T_f, w_1') - R(T_f, w_2')}{8}\right]$$

PROOF The connection between the Reidemeister torsion of the mapping torus and the Lefschetz zeta function follows from formula (6.8):

$$\tau(T_f; p^*(\det \partial_{w_1} / \det \partial_{w_2})) = \mid L_f(\lambda) \mid^{-1},$$

where

$$\lambda = hol(g; \det \partial_{w_1} / \det \partial_{w_2})$$

The result now follows from the previous theorem 65.

6.3.4   The Reidemeister torsion and theta-multipliers

We recall some well known properties of theta-functions (see [54]). Fix an integer $g \geq 1$. A characteristic $\mathbf{m}$ is a row vector $\mathbf{m} = (m_1^*, ..., m_g^*, m_1^{**}, ..., m_g^{**})$ each of whose entries is zero or one. The parity $e(\mathbf{m}) = m_1^* \cdot m_1^{**} + ... + m_g^* \cdot m_g^{**} \in \mathbb{Z}/2\mathbb{Z}$ of the characteristic $\mathbf{m}$ is said to be even (odd) when $e(\mathbf{m}) = 0$ ($= 1$).

Let S_g denote the Siegel space of degree g. We shall write elements $T \in Sp_{2g}(\mathbb{Z})$ as block matrices:

$$T = \begin{pmatrix} A & B \\ C & D \end{pmatrix}.$$

The theta function with characteristic $\mathbf{m}$ is a function $\theta_\mathbf{m} : S_g \times \mathbb{C}^g \to \mathbb{C}$ satisfying the following transformations law [54] :

$$\theta_{T \cdot \mathbf{m}}(z(C\tau + D)^{-1}, (A\tau + B)(C\tau + D)^{-1}) =$$

$$= \gamma_\mathbf{m}(T) \cdot \det(C\tau + D)^{1/2} \cdot \exp\left(\pi i z(C\tau + D)^{-1}\, {}^t z\right) \cdot \theta_\mathbf{m}(\tau, z)$$

where

$$T \cdot \mathbf{m} = \mathbf{m} \cdot T^{-1} + ((D\,{}^t C)_0 (B\,{}^t A)_0) \quad (\text{mod } 2)$$

and $(\ )_0$ denotes the row vector obtained by taking the diagonal elements of the matrix. The number $\gamma_\mathbf{m}(T)$ is an eighth root of unity, and are known as the theta multiplier of T for the characteristic $\mathbf{m}$. The action of the symplectic group $Sp_{2g}(\mathbb{Z})$ on the characteristic (denoted above $\mathbf{m} \mapsto T.\mathbf{m}$) preserves parity, and is transitive on characteristics of a given parity. We let $\Gamma(\mathbf{m}) = \{T \in Sp_{2g}(\mathbb{Z}) \mid T \cdot \mathbf{m} = \mathbf{m}\}$. If Γ_2 is the principal congruence subgroup of level 2 in $Sp_{2g}(\mathbb{Z})$, i.e.

$$\Gamma_2 = \{T \in Sp_{2g}(\mathbb{Z}) \mid T \equiv I \quad (\text{mod } 2)\},$$

then $\Gamma_2 \subset \Gamma(\mathbf{m})$ for every $\mathbf{m}$.

For a diffeomorphism $f : V^2 \to V^2$ of a Riemann surface V^2 of genus g, the induced map on the homology $T = f_* : H_1(V^2, \mathbb{Z}) \to H_1(V^2, \mathbb{Z})$ is an element in the integral symplectic group $Sp_{2g}(\mathbb{Z})$. There is a bijection between the set of Spin structures on V^2 and the set of characteristics $\mathbf{m}$. We shall let $w_\mathbf{m}$ denote the spin structure corresponding to the characteristic $\mathbf{m}$. Thus a diffeomorphism $f : V^2 \to V^2$ preserves the spin structure $w_\mathbf{m}$ iff

$f_* : H_1(V, \mathbb{Z}) \to H_1(V, \mathbb{Z})$ is in $\Gamma(\mathbf{m})$. If f_* is in Γ_2 then f preserves each of the spin structures on V^2. Let T be any element of $\Gamma_2 \subset Sp_{2g}(\mathbb{Z})$ and $f : V^2 \to V^2$ be a diffeomorphism with $f_* = T : H_1(V^2, \mathbb{Z})) \to H_1(V^2, \mathbb{Z})$. Consider again the bundle $p : T_f \to S^1$ over the circle S^1 with the fiber V^2. Let $w_{\mathbf{m}}$ and $w_{\mathbf{n}}$ be two spin structures on V^2 corresponding to any two *even* characteristic $\mathbf{m}$ and $\mathbf{n}$. We have a flat complex determinant line bundle $\det \partial_{w_{\mathbf{m}}} / \det \partial_{w_{\mathbf{n}}}$ over the base S^1 and its pullback $p^*(\det \partial_{w_{\mathbf{m}}} / \det \partial_{w_{\mathbf{n}}})$ over the mapping tori T_f.

Theorem 67

$$\tau(T_f; p^*(\det \partial_{w_{\mathbf{m}}} / \det \partial_{w_{\mathbf{n}}}) = \mid L_f(\gamma_{\mathbf{m}}(T)/\gamma_{\mathbf{n}}(T)) \mid^{-1}.$$

Proof Lee,Miller and Weintraub [61] proved that

$$\gamma_{\mathbf{m}}(T)/\gamma_{\mathbf{n}}(T) = \exp\left(-2\pi i \frac{R(T_f, w_{\mathbf{m}}) - R(T_f, w_{\mathbf{n}})}{8}\right)$$

Now the statement of the theorem it follows from theorem 66.

6.4 Topology of an attraction domain and the Reidemeister torsion

6.4.1 Introduction

Assume that on a smooth compact manifold M of dimension n there is given a tangential vector field X of class C^1, and consider the corresponding system of differential equations

$$\frac{dx}{dt} = X(x), \tag{6.11}$$

Let $\phi(t, x)$ is the trajectory of (1) passing through the point x for $t = 0$. We shall say that the set I is the attractor or the asymptotically stable compact invariant set for system (1) if for any neighborhood U of I there is the neighborhood W, $I \subset W \subset U$ such that
1) for any $x \in W$ $\phi(t, x) \in U$, if $t \in [0, +\infty)$,

2) for any $x \in W$ $\phi(t, x) \longrightarrow I$, when $t \longrightarrow +\infty$.

By a Lyapunov function $V(x)$ for attractor I we mean a function that satisfies following conditions

1)$V(x) \in C^1(U - I), \quad V(x) \in C(U),$

2)$V(x) > 0, x \in U - I; \quad V(x) = 0, x \in I,$

3) The derivative by virtue of the system (1) $\frac{dV(x)}{dt} > 0$ in $U - I$.

Such Lyapunov function $V(x)$ for I always exist [100]. Suppose that S is a level surface of Lyapunov function $V(x)$ in U.The conditions 3) and the Implicit Function Theorem imply that the level surface S is a compact smooth $n - 1$-dimensional manifold transverse to the trajectories of (1) and trajectories of (1) intersect S on the descending side of the Lyapunov function $V(x)$. Any two level surfaces of the Lyapunov function $V(x)$ are diffeomorphic.Note that manifold S is determined up to diffeomorphism by the behavior of trajectories of the system (1) in $U - I$ and does not depend on the choice of the Lyapunov function $V(x)$ and its level. Let $N \supset I, \dim N = n$, be a compact smooth manifold with the boundary $\partial N = S$.

In this article we will study the dependence of the topology of the attraction domain

$$D = \{x \in M - I : \quad \phi(t, x) \longrightarrow I, \text{when } t \longrightarrow +\infty\}$$

of the attractor I and of the level surface S of the Lyapunov function $V(x)$ on the dynamical properties of the system (1) on the attractor. The investigation of the topological structure of the level surfaces of the Lyapunov function was initiated by Wilson [100]. Note that the attraction domain D is diffeomorphic to $S \times R^1$,since each trajectory of system (1) in the invariant attraction domain D intersects $n - 1$-dimensional manifold S exactly once.Hence it follows that the homology groups of D and S are isomorphic.

6.4.2   Morse-Smale systems

We assume in this section that system (1) is given in R^n and is a Morse-Smale system on manifold N, i.e. the following conditions are satisfied:
1) A set of nonwandering trajectories Ω of system (1) is the union of a finite number of hyperbolic stationary points and hyperbolic closed trajectories,
2) Stable and unstable manifolds of stationary points and closed trajectories intersect transversally.

A stable or unstable manifold of a stationary point or a closed trajectory p is denoted by $W^s(p)$ and $W^u(p)$. Let a^k be the number of stationary points p of system (1) in I such that $\dim W^u(p) = k$, b_k is the number closed orbits q of system (1) in I such that $\dim W^u(q) = k$, $M_k = a_k + b_k + b_{k+1}$, $B_k = \dim H_k(D; Q) = \dim H_k(S; Q)$, $\chi(D) = \chi(S)$ is the Euler characteristic of D and S.

Theorem 68 *The numbers B_k and M_k satisfy the following inequalities:*

$$B_0 \leq M_0 + M_{n-1} - M_n,$$

$$B_1 - B_0 \leq M_1 - M_0 + M_{n-2} - M_{n-1} + M_n,$$

$$B_2 - B_1 + B_0 \leq M_2 - M_1 + M_0 + M_{n-3} - M_{n-2} + M_{n-1} - M_n, \quad (6.12)$$

$$\dotfill,$$

$$\sum_{i=0}^{n-1}(-1)^i \cdot B_i = \chi(S) = \chi(D) = (1 + (-1)^{n-1})\sum_{i=0}^{n}(-1)^i \cdot M_i.$$

To prove the theorem we need several preliminary definitions and lemmas.

Lemma 36
$$B_r = B_r(N) + B_{n-1-r}(N), \quad (6.13)$$
where $B_r(N) = \dim H_r(N; Q)$.

PROOF It is possible to assume that S lies on the n-dimensional sphere S^n. Suppose $C = Cl(S^n - N)$ is the closure of the complement of N. Then $S = C \cap N$. The manifolds C and N are compact and intersect only on the boundary. Consequently, we have the exact reduced Maier-Vietoris sequence :

$$.... \to \tilde{H}_{r+1}(S^n; Q) \to \tilde{H}_r(S; Q) \to \tilde{H}_r(N; Q) \oplus \tilde{H}_r(C; Q) \to \tilde{H}_r(S^n; Q) \to ...$$

Set $1 \leq r < n - 1$. Then $\tilde{H}_r(S^n; Q) = \tilde{H}_{r+}(S^n; Q) = 0$, and therefore

$$\tilde{H}_r(S; Q) = \tilde{H}_r(N; Q) \oplus \tilde{H}_r(C; Q).$$

Hence $\tilde{B}_r(S) = \tilde{B}_r(N) + \tilde{B}_r(C)$, where $\tilde{B}_r$ denotes the dimension of $\tilde{H}_r$. From Alexander duality it follows that $\tilde{B}_r(C) = \tilde{B}_r(N - \partial N)$ It is easy to show that N and $N - \partial N$ are homotopically equivalent(for example, by means of the collar theorem). Therefore $\tilde{B}_r(N - \partial N) = \tilde{B}_r(N)$, hence

$$\tilde{B}_r(S) = \tilde{B}_r(N) + \tilde{B}_{n-1-r}(N) \text{ and } B_r(S) = B_r(N) + B_{n-1-r}(N).$$

The manifold S is closed and orientable ; therefore in case $r = n - 1$ we obtain the following from Poincare duality:

$$B_0(S) = B_0(N) + B_{n-1}(N).$$

Thus the lemma is proved for all $r = 0, 1, 2, ..., n - 1$. We make the following definition:

Definition 16 *The stationary point p of the vector field X with $\dim W^u(p) = k$ has standard form if there exist local coordinates $x_1, x_2, ..., x_k; y_1, y_2, .., y_{n-k}$ in some neighborhood of the point p such that*

$$X = x_1 \frac{\partial}{\partial x_1} + ... + x_k \frac{\partial}{\partial x_k} - y_1 \frac{\partial}{\partial y_1} - .... - y_{n-k} \frac{\partial}{\partial y_{n-k}}$$

in this neighborhood.

The standard form for a closed trajectory is defined analogously [38].

Lemma 37 *[38] If X_0 is a Morse-Smale vector field on the manifold N, then there exists a path in the space of smooth vector fields on N, X_t, $t \in [0,1]$, such that:*
1) X_t is a Morse-Smale vector field for all $t \in [0,1]$;
2) the stationary points and closed trajectories of the field X_1 are all in standard form.
By using this result we replace the original vector field by a vector field for which all stationary points and closed trajectories have standard form.

Lemma 38 *[38] Suppose X is a Morse-Smale vector field on an orientable manifold, γ is a closed trajectory in standard form, $\dim W^u(\gamma) = k + 1$, U is a sufficiently small neighborhood of γ. There exists a Morse-Smale vector field Y which coincides with x outside of U, in U has stationary points p and q, and has no other stationary points or closed orbits. Moreover $\dim W^u(p) = k, \dim W^u(q) = k + 1$.*

By means of this lemma we replace a vector field on the manifold N by the field Y having no closed trajectories.In this connection the number of stationary points of the field Y with $\dim W^u(p) = k$ will be equal to $a_k + b_k + b_{k+1}$

Lemma 39 *[88] Suppose Y is a smooth Morse-Smale vector field without closed trajectories on a compact manifold N with boundary, the stationary points of Y have standard form, and the vector field on ∂N is directed inward. Then there exists a Morse function f on N such that:*
1) the critical points of the function f coincide with stationary points of the field Y, the index of a critical point of the function f coincides with the dimension of the unstable manifold of this point ;
2) if $p \in N$ is a critical point of f then $f(p) = \dim W^u(p)$;
3) $f(\partial N) = \frac{n+1}{2}$.

Lemma 40 *[88] Suppose f is a Morse function on N satisfying conditions 1) - 3). Then:*

$$B_0(N) \leq M_0,$$
$$B_1(N) - B_0(N) \leq M_1 - M_0,$$
$$B_2(N) - B_1(N) + B_0(N) \leq M_2 - M_1 + M_0, \qquad (6.14)$$

$$\sum_{i=0}^{n}(-1)^i \cdot B_i(N) = \chi(N) = \sum_{i=0}^{n}(-1)^i \cdot M_i.$$

Corollary 25 *The following inequalities hold:*

$$B_{n-1}(N) \leq M_{n-1} - M_n,$$
$$B_{n-2}(N) - B_{n-1}(N) \leq M_{n-2} - M_{n-1} + M_n,$$
$$B_{n-3}(N) - B_{n-2}(N) + B_{n-1}(N) \leq M_{n-3} - M_{n-2} + M_{n-1} - M_n, \qquad (6.15)$$

$$\sum_{i=0}^{n-1}(-1)^i \cdot B_i(N) = \chi(N) = (1 + (-1)^{n-1})\sum_{i=0}^{n-1}(-1)^i \cdot M_i.$$

PROOF OF THE COROLLARY. From the fact that $H_n(N) = 0$ it follows that

$$-B_{n-1}(N) + B_{n-2}(N)... + (-1)^n B_0(N) = M_n - M_{n-1}... + (-1)^n M_0 \quad (6.16)$$

Adding (6.16) to the inequalities (6.14) and subtracting (6.16) from the inequalities (6.14) we obtain system of the inequalities (6.15).

PROOF OF THEOREM 68 From lemma 1, inequalities (6.14) and (6.15) it follows that

$$B_k - B_{k-1} + B_{k-2} - ... + (-1)^k B_0 = B_k(N) - B_{k-1}(N) + ... + (-1)^k B_0(N) +$$

$$+B_{n-1-k}(N) - B_{n-k}(N) + B_{n-k+1}(N) + ... + (-1)^k B_{n-1}(N) \leq$$

$$\leq M_k - M_{k-1} + ... + (-1)^k M_0 + M_{n-1-k} - M_{n-k} + ... + (-1)^k M_{n-1} - (-1)^k M_n;$$

$$\chi(S) = \sum_{k=0}^{n-1} (-1)^k \cdot B_k = \sum_{k=0}^{n-1} (-1)^k (B_k(N) + B_{n-1-k}(N)).$$

Using the fact that $H_n(N) = 0$ and changing the index of summation we obtain

$$\chi(S) = (1 + (-1)^{n-1}) \sum_{i=0}^{n} (-1)^i \cdot B_i(N) = (1 + (-1)^{n-1}) \sum_{i=0}^{n} (-1)^i \cdot M_i$$

The theorem is proved.

6.4.3 A formula for the Euler characteristic

The last identity in Theorem 68 is also true in more general situation. Namely, assume that system (6.11) is an autonomous system of differential equations having a finite number of stationary points in attractor I. Denote by Index (p) the indices of the vector field X at stationary point p.

Theorem 69

$$\chi(D) = \chi(S) = ((-1)^n - 1) \cdot \sum_{p \in I} \text{Index } (p). \quad (6.17)$$

PROOF The vector field X is directed on ∂N into N . Therefore from the Poincare-Hopf theorem [65], by replacing X by $-X$ we obtain

$$\chi(N) = (-1)^n \cdot \sum_{p \in I} ind(p).$$

It is known that $\chi(\partial N) = (1 + (-1)^{n-1}) \cdot \chi(N)$. Hence

$$\chi(D) = \chi(S) = ((-1)^n - 1) \cdot \sum_{p \in I} \text{Index } (p).$$

Corollary 26 *Suppose the stationary points on I are hyperbolic , a_k is the number of stationary points of I with $\dim W^u(p) = k$. Since for a hyperbolic stationary point p with $\dim W^u(p) = k$ the index of the vector field at it is equal to $(-1)^k$, we obtain the following formula:*

$$\chi(D) = \chi(S) = ((-1)^n - 1) \cdot \sum_{k=0}^{n} (-1)^k \cdot a_k. \qquad (6.18)$$

For $n = 3$ S is union of finite number of spheres with handles. Suppose m is the number of connected components, p is the total number of handles of the manifold S. Then $\chi(S) = 2m - 2p$. Hence we obtain

Corollary 27

$$m - p = -\sum_{p \in I} \text{Index } (p). \qquad (6.19)$$

In the case, when stationary points are hyperbolic

$$m - p = a_o - a_1 + a_2 - a_3.$$

6.4.4 The Reidemeister torsion of the level surface of a Lyapunov function and of the attraction domain of the attractor

In this section we consider the flow (6.11) with circular chain recurrent set $R \subset I$. The Reidemeister torsion of the attraction domain D and of the level surface S is the relevant topological invariant of D and S which is calculated in theorem 70 and corollary 28 via closed orbits of flow (6.11) in the attractor I.

The point $x \in M$ is called chain-recurrent for flow (6.11) if for any $\varepsilon > 0$ there exist points $x_1 = x, x_2, ..., x_n = x$ and real numbers $t(i) \geq 1$ such that $\rho(\phi(t(i), x_i), x_{i+1}) < \varepsilon$ for $1 \leq i < n$. Let $R \subset I$ be a set of chain-recurrent points of equation (6.11) on the manifold N defined above. We assume in this section that R is circular, i.e. there is a smooth map $\theta : U \to R^1/Z, U$ a neighborhood of R in N, on which $\frac{d}{dt}(\theta \circ \phi(t, x)) > 0$. In other words, there is a cross-section of the flow (6.11) on R, namely, a level set of θ on $int(U)$. For instance, if R is finite i.e., consists of finitely many closed orbits, then R is circular. More generally, if ϕ on R has no stationary points and the topological dimension of R is 1, then R is circular. For example ,if ϕ is a nonsingular Smale flow , so that R is hyperbolic and 1-dimensional, then R is circular. If $U \in N$ is such that $\cap_{t \in R^1} \phi_t(U) = J$ is compact and $J \in intU$, then we say that U is an isolating neighborhood of the isolated invariant set J. According to Conley [14], there is a continuous function $G :\to R^1$ such that G is decreasing on $N - R$ and $G(R)$ is nowhere dense in R^1. Taking an open neighborhood W of $G(R)$ and $U = G^{-1}(W)$, we see that U is an isolating neighborhood for some isolating invariant set J and that $J \to R$ as $W \to G(R)$ [40].This proves that the chain recurrent set R can be approximated by the isolated invariant set J .In particular we can make J circular. Further, there are finitely many points $x_i < x_{i+1}$ in $R^1 - G(R)$ such that $G^{-1}[x_i, x_{i+1}]$ isolates an invariant set J_i so that $J = \cup J_i$ is as closes as we like to R . In particular, we can make J circular. For the sequel we need the isolating blocks [14]. A compact isolating neighborhood B_i of J_i is said to be isolating block if:

1)B_i is smooth manifold with corners,

2) $\partial B_i = b_i^+ \cup b_i^- \cup b_i^0$ where each term is compact manifold with boundary,

3) the trajectories $\phi(t, x)$ is tangent to b_i^0 and $\partial b_i^0 = (b_i^+ \cup b_i^-) \cap b_i^0$,

4) the trajectories $\phi(t, x)$ is transverse to b_i^+ and b_i^-, enters B_i on b_i^+ and exists on b_i^-.

Since J_i is circular there is a smooth map $\theta_i : B_i \to R^1/Z$ such that $\frac{d}{dt}(\theta_i \circ \phi(t, x)) > 0$ on B_i.By perturbing θ_i, we can make θ_i transverse to $0 \in R^1/Z$ on $B_i, b_i^+, b_i^-, b_i^0, \partial b_i^+, \partial b_i^-, \partial b_i^0$. Now let $Y_i = \theta_i^{-1}(0) \cup b_i^-, Z_i = b_i^-$. Then (Y_i, Z_i) is a simplicial pair . We define a continuous map $r_i : Y_i \to Y_i$ as follows. If $y_i \notin Z_i$ then $r_i(y) = \phi(\tau, y)$ where $\tau = \tau(y) > 0$ is the smallest positive time t for which $\phi(t, y) \in Y_i$. Since $\phi(t, x)$ is transverse out on $b_i^- 0 Z_i$, we see that $\tau(y)$ is near 0 for y near Z_i.Thus τ extends continuously to Z_i if we set $\tau/Z_i = 0$. Now let E be a flat complex vector bundle of finite dimension on B_i. There is a bundle map $\alpha_i : r_i^*(E|Y_i) \to E|Y_i$ defined by pulling back along trajectory from y to $r_i(y)$, using the flat connection on E. This determines an endomorphism

$$(\alpha_i)_* : H^*(Y_i, Z_i; r_i^*E) \to H^*(Y_i, Z_i; E). \qquad (6.20)$$

Since there is a natural induced map $r_i^* : H^*(Y_i, Z_i; E) \to H^*(Y_i, Z_i; r_i^*E)$ we obtain the endomorphism

$$\beta_i = (\alpha_i)_* \cdot r_i^* : H^*(Y_i, Z_i; E) \to H^*(Y_i, Z_i; E). \qquad (6.21)$$

So the relative Lefschetz number

$$L(\beta_i) = \sum_{k=o}^{n-1} (-1)^k \cdot \mathrm{Tr}\ (\beta_i)_k \qquad (6.22)$$

is defined. According to Atiyah and Bott [1] the numbers $L(\beta_i)$ can be computed from the fixed point set of r_i in $Y_i - Z_i$. If Fix $(r_i) - Z_i$ is a finite set of points p with the Lefschetz index Index $_L(r_i, p)$ and $(\alpha_i)_p : E_p \to E_p$ is the endomorphism of the fiber at p, then one has the relative Lefschetz formula

$$L(\beta_i) = \sum_p \mathrm{Index}\ _L(r_i, p) \cdot \mathrm{Tr}\ (\alpha_i)_p. \qquad (6.23)$$

We see that $L(\beta_i^n), n \geq 1$, counts the periodic points of period n for β_i which are not in Z_i, i.e. the closed orbits of system (6.11) that wrap n times around R^1/Z under θ_i, with a weight coming from the holonomy of E around these closed orbits. Now , consider the twisted Lefschetz zeta function [40] for E and (B_i, b_i^-):

$$L_i(z) \equiv L_{\beta_i}(z) := \exp\left(\sum_{n=1}^{\infty} \frac{L(\beta_i^n)}{n} z^n\right) \qquad (6.24)$$

we now turn to the R-torsion of pairs. Suppose that L is a CW-subcomplex of K and consider the relative cochain complex

$$C^*(K, L; E) = ker(C^*(K; E) \to C^*(L; E|_L)).$$

Then one has a natural isomorphism $|C^*(K, L; E)| \cong \otimes_j |V|$, where j runs over the i-cells in $K \backslash L$. So our flat density on E gives a density Δ_i on the relative i-cochains in E. Thus we again have an R-torsion denoted $\tau(K, L; E, D_i)$

for some choice of positive densities D_i on $H^i(K, L; E)$. If $H^i(K, L; E) = 0$, we say that E is acyclic for (K, L) and we simply write $\tau(K, L; E)$, when the D_i are chosen standard. Let $\rho_E : \pi_1(N, p) \to GL(E_p)$ is holonomy representation for acyclic bundle E on orientable manifold N, $\dim N = n$, ρ_E^* is the cotragredient representation of ρ_E and E^* is a flat complex vector bundle with the holonomy ρ_E^*. We suppose that $\det \rho_E = 1$. Let $L_i^*(z)$ is the twisted Lefschetz zeta function for E^* and (B_i, b_i^-), and

$$L^*(z) = \prod_i L_i^*(z), \qquad L(z) = \prod_i L_i(z). \tag{6.25}$$

Theorem 70

$$\tau(D; E) = \tau(S; E) = |L(1)|^{-1} \cdot |L^*(1)|^{\varepsilon(n)}, \tag{6.26}$$

where $\varepsilon(n) = (-1)^n$.

PROOF Consider the function G. Smoothing the level set $G^{-1}(x_i)$ by sliding it along the flow, one obtains a smooth region $N_i \subset N$ with

$$G^{-1}((-\infty, x_i - \varepsilon)) \subset G^{-1}((-\infty, x_i + \varepsilon))$$

, such that the trajectories $\phi(t, x)$ transverse to ∂N_i, for large i we have $N_i = N$ and $\partial N^- = \emptyset$. If ε is small one has that $N_{i+1} - N_i$ isolates J_i. Then by properties of the Reidemeister torsion [40] one finds:

$$\tau(N; E) = \prod_i \tau(N_{i+1}, N_i; E) = \prod_i \tau(B_i, b_i; E) \tag{6.27}$$

D.Fried proved [40] that E is acyclic for (B_i, b_i^-) iff $I - \beta_i$ is invertible and then

$$\tau(B_i, b_i; E) = |L_i(z)|^{-1}|_{z=1} \tag{6.28}$$

So we have

$$\tau(N; E) = \prod_i |L_i(1)|^{-1} = |L(1)|^{-1} \tag{6.29}$$

From the multiplicative law (6.1) for the Reidemeister torsion it follows:

$$\tau(N; E) = \tau(N, \partial N = S; E) \cdot \tau(\partial N = S; E) \tag{6.30}$$

Using Milnor's duality theorem for the Reidemeister torsion [65] we have:

$$\tau(N, \partial N = S; E) = \tau(N; E^*)^{(-1)^n} \qquad (6.31)$$

From formula (6.29) it follows that

$$\tau(N; E^*) = \prod_i |L_i^*(1)|^{-1} = |L^*(1)|^{-1} \qquad (6.32)$$

Since the attraction domain D is diffeomorphic to $S \times R^1$ then the Reidemeister torsion $\tau(D; E) = \tau(S; E)$ by the simple homotopy invariance of the Reidemeister torsion. Now from (6.29), (6.30), (6.31), (6.32) we have:

$$\tau(D; E) = \tau(S = \partial N; E) = \tau(N; E) \cdot \tau^{-1}(N, \partial N = S; E) =$$

$$= \tau(N; E) \cdot \tau(N; E^*)^{(-1)^{n+1}} = |L(1)|^{-1} \cdot |L^*(1)|^{(-1)^n}$$

Suppose now that the system(6.11) on the manifold N is a nonsingular almost Morse-Smale system. This means that (6.11) has finitely many hyperbolic prime periodic orbits γ and no other chain-recurrent points. Over the orbit γ lies a strong unstable bundle $E^u(\gamma)$ of some dimension $u(\gamma)$. Let $\delta(\gamma)$ be $+1$ if E^u is orientable and -1 if it is not. Let $\varepsilon(\gamma) = (-1)^{u(\gamma)}$

Corollary 28
$$\tau(D; E) = \tau(S = \partial N; E) =$$
$$= \prod_\gamma |\det(I - \delta(\gamma) \cdot \rho_E(\gamma))|^{\varepsilon(\gamma)} \times (\prod_\gamma |\det(I - \delta(\gamma) \cdot \rho_E^*(\gamma))|^{\varepsilon(\gamma)})^{(-1)^{n+1}}$$

PROOF According to D.Fried [40] if J_i is a prime hyperbolic closed orbit γ then
$$|L_i(1)|^{-1} = |\det(I - \delta(\gamma) \cdot \rho_E(\gamma))|^{\varepsilon(\gamma)}$$

Now, the statement it follows from theorem 70.

6.5 Integrable Hamiltonian systems and the Reidemeister torsion

Let M be a four-dimensional smooth symplectic manifold and the system (6.11) be a Hamiltonian system with a smooth Hamiltonian H. In the Darboux coordinates such system has the form:

$$\frac{dp_i}{dt} = \frac{\partial H}{\partial q_i} \qquad\qquad (6.33)$$

$$\frac{dq_i}{dt} = \frac{\partial H}{\partial p_i}.$$

As the Hamiltonian H is the integral of the system (6.33), then three-dimensional level surface $Q = [H = const]$ is invariant for the system (6.33). The surface Q is called the isoenergetic surface or the constant -energy surface. Since M is orientable(as a symplectic manifold), the surface Q is automatically orientable in all cases. Suppose that the system (6.33) is complete integrable (in Liouville's sense) on the surface Q. This means, that there is the smooth function f(the second integral), which is independent with H and for the Poisson bracket $[H, f] = 0$in the neighborhood of Q. We shall call the integral f a Bott integral on the isoenergetic surface Q, if its critical points form critical nondegenerate smooth submanifolds in Q.This means that Hessian $d^2 f$ of the function f is nondegenerate on the planes normal to the critical submanifolds of the function f. A.T. Fomenko [37] proved that a Bott integral on compact nonsingular isoenergetic surface Q can have only three types of critical submanifolds: circles, tori, Klein bottles. The investigation of the concrete systems shows [37] that it is a typical situation when the integral on Q is a Bott integral. In the classical integrable cases of the solid body movement(cases of the Kovalevskaya, Goryachev-Chaplygin, Clebsch, Manakov) the Bott integrals are a round Morse functions on the isoenergetic surfaces. The round Morse function is a Bott function all whose critical manifolds are circles. Note that critical circles of f is a periodic solution of the system(6.33) and the number of this circles is finite. Suppose now that the Bott integral f is a round Morse function on the closed isoenergetic surface Q. Let us recall the concept of the separatrix diagram of the critical circle γ for a Bott function f. Let $x \in \gamma$ be an arbitrary point and $N_X(\gamma)$ be a disc of small radius normal to γ at x. The restriction of f to the $N_X(\gamma)$ is a normal Morse function with the critical point x having a certain index $\lambda = o, 1, 2$. The separatrix of the critical point x is the integral trajectory of the field $\mathrm{grad} f$, which is entering or leaving x. The union of all the separatrices entering the point x gives a disc of dimension λ and is called the incoming separatrix diagram(disc). The union of outgoing separatrices gives a disc of additional dimension and is called the outgoing separatrix diagram(disc). Varing the point x and constructing the incoming and outgoing separatrix discs for each point x, we obtain the incoming and outgoing separatrix diagrams of circleγ. Let $u(\gamma)$ be the dimensi on of the outgoing separatrix diagram of γ, and $\delta(\gamma)$ be +1 if this separatrix is orientable , and -1 if it is not. Let $\varepsilon(\gamma) = (-1)^{u(\gamma)}$. Suppose that $\rho_E : \pi_1(Q, p) \to GL(E_p)$ is holonomy representation for acyclic bundle E over Q; E_p is a fiber at point p.

Theorem 71

$$\tau(Q; E) = \prod_\gamma |\det(I - \delta(\gamma) \cdot \rho_E(\gamma))|^{\varepsilon(\gamma)} \qquad (6.34)$$

provided that each determinant occuring in this formula is nonzero.

PROOF As a Bott integral f on Q is a round Morse function then according to Thurston [92] Q has a round handle decomposition whose core circles are critical circles of f. According to Asimov [6] if Q has a round handle decomposition then Q has a nonsingular Morse-Smale flow whose closed orbits are exactly the core circles of the round handles, i.e. critical circles γ of f. Consider this nonsingular Morse-Smale flow. For a closed orbit γ of such flow the dimension $u(\gamma)$ of unstable manifold $W^u(\gamma)$ is exactly the dimension of the outgoing separatrix diagram of γ as critical circle of f. One can choose as isolating block B for γ to be a bundle over S^1 with fiber F a simplicial disc so that $F \cong I^u \times I^s, u = u(\gamma), s = 4 - u(\gamma)$, and $F \cap b^- \cong \partial I^u \times I^s \subset F$. By collapsing I^s to a point one can produce a simple homotopy equivalence of (B, b^-) to $(X_\gamma, \partial X_\gamma)$ where X_γ is the unstable disc bundle over γ. Give Q its Smale filtration by compact submanifolds Q_i of top dimension so that $Q_i \subset int Q_{i+1}, Q_0 = \emptyset, Q_i = Q$ for large i , flow is transverse inward on ∂Q_i and $Q_{i+1} - Q_i$ is an isolating neighborhood for a hyperbolic closed orbit γ.Then (Q_{i+1}, Q_i) has the same homotopy type as $(X_\gamma, \partial X_\gamma)$ and by properties of the Reidemeister torsion [40] one finds:

$$\tau(Q; E) = \prod_i \tau(Q_{i+1}, Q_i; E) = \prod_\gamma \tau(X_\gamma, \partial X_\gamma; E) =$$

$$= \prod_\gamma |\det(I - \delta(\gamma) \cdot \rho_E(\gamma))|^{\varepsilon(\gamma)}.$$

Bibliography

[1] M.Atiyah, R. Bott, A Lefschetz fixed point formula for elliptic complexes II. Annals of Math., v.88(1968), 451-491.

[2] A.Aigner, Combinatorial Theory. Springer, Heidelberg , 1979.

[3] D.Anderson, E.Brown, E.Peterson, The structure of the spin cobordism ring. Annals of Math., 86 (1967), 271-298.

[4] D. Anosov, The Nielsen number of maps of nilmanifolds. Russian Math. Surveys , 40 (1985), 149-150.

[5] M. Artin and B. Mazur, On periodic points, Annals of Math., 81 (1965), 82-99.

[6] D. Asimov, Round handle and non-singular Morse-Smale flows. Annals of Math., v.102(1975), 41-54.

[7] M. Atiyah, V. Patodi, I. Singer, Spectral asymmetry and Riemann geometry I. Math. Proc. Cambr. Phil. Soc., 77 (1975), 43-69.

[8] I. Babenko, S. Bogatyi, Private communication.

[9] J. Bismut, D. Freed, The analysis of elliptic families. II. Dirac families, eta invariants and the holonomy theorem. Comm. Math. Phys., 107 (1986), 103-163.

[10] J. S. Birman, Braids, links and mapping class groups. Ann. Math. Studies vol. 82, Princeton Univ. Press, Princeton, 1974.

[11] R. Bowen and O. Lanford, Zeta functions of restrictions of the shift transformation. Proc. Global Anal., 1968, 43-49.

[12] R.Brooks, R.F.Brown, J.Pak, and D.H.Taylor, Nielsen numbers of maps of tori. Proc. Am. Math. Soc., 52 (1975), 398-400.

[13] J. Cheeger, Analytic torsion and the heat equation. Annals of Math., 109 (1979), 259-322.

[14] C.Conley, Isolated invariant sets and the Morse index. C. B. M. S. Reg. Conf. ser., v.38(1978).

[15] P. Deligne, La conjecture de Weil. Publ. math. IHES, 43(1974), 273-307.

[16] A. Dold, Fixed point indices of iterated maps. Inventiones Math. 74 (1983), 419-435.

[17] Encyclopedic Dictionary of Mathematics. MIT Press, Cambridge, Mass., 1977.

[18] J. Eichhorn, Elliptic operators on noncompact manifolds. Teubner-Texte Math. 106(1988), 4-169.

[19] C. Epstein, The spectral theory of geometrically periodic hyperbolic 3-manifolds. Memoirs of the AMS, vol. 58, number 335, 1985.

[20] E. Fadell, S. Husseini, The Nielsen number on surfaces, Topological methods in nonlinear functional analysis. Contemp. Math. vol. 21, AMS, Providence, 1983, 59-98.

[21] E. Fadell and S. Husseini, On a theorem of Anosov on Nielsen numbers for nilmanifolds, Nonlinear Functional Anal. Appl. 173 (1986), 47-53.

[22] A. Fathi and M. Shub, Some dynamics of pseudo-Anosov diffeomorphisms. Asterisque 66-67 (1979), 181-207.

[23] A. L. Fel'shtyn, New zeta function in dynamic. in Tenth Internat. Conf. on Nonlinear Oscillations, Varna, Abstracts of Papers, Bulgar. Acad. Sci., 1984, 208

[24] A.L. Fel'shtyn, New zeta functions for dynamical systems and Nielsen fixed point theory. in : Lecture Notes in Math. 1346, Springer, 1988, 33-55.

[25] A.L.Fel'shtyn, The Reidemeister zeta function and the computation of the Nielsen zeta function. Colloquium Mathematicum 62 (1) (1991), 153-166.

[26] A.L.Fel'shtyn, The Reidemeister, Nielsen zeta functions and the Reidemeister torsion in dynamical systems theory. Zap. nauch. sem. Leningrad. otd. Math. Inst. Steklowa. Geometry and Topology 1,V.193, 1991,p.119-142.

[27] A.L. Fel'shtyn, Zeta functions in Nielsen theory. Functsional Anal. i Prilozhen 22 (1) (1988), 87-88 (in Russian); English transl.: Functional Anal. Appl. 22 (1988), 76-77.

[28] A.L. Fel'shtyn, A new zeta function in Nielsen theory and the universal product formula for dynamic zeta functions. Functsional Anal. i Prilozhen 21 (2) (1987), 90-91 (in Russian); English transl.: Functional Anal. Appl. 21 (1987), 168-170.

[29] A.L. Fel'shtyn, The Rochlin invariant, eta-invariant and Reidemeister torsion. St.Petersburg Mathematical Journal v.3,n.4,1991, p.197-206.

[30] A.L. Fel'shtyn, Attractors, integrable hamiltonian systems and the Reidemeister torsion. Progress in Nonlinear Differential Equations and Their Applications, v.12,1994,p.227-234. Birkhäuser.

[31] A.L.Fel'shtyn,R.Hill, Dynamical zeta functions, Nielsen theory and Reidemeister torsion. Mathematica Göttingensis.Heft 22.August 1992, 36 p.

[32] A.L.Fel'shtyn,R.Hill, Dynamical zeta functions, Nielsen theory and Reidemeister torsion. Contemporary Mathematics, v.152, 1993,p.43-69.

[33] A.L.Fel'shtyn,R.Hill, The Reidemeister zeta function with applications to Nielsen theory and a connection with Reidemeister torsion. K-theory,v.8,n.4,1994,p.367-393 .

[34] A.L.Fel'shtyn,R.Hill, Trace formulae, zeta functions, congruences and Reidemeister torsion in Nielsen theory. Forum Mathematicum, v.10, n.6, 1998, 641-663.

[35] A.L.Fel'shtyn,R.Hill, Congruences for Reidemeister and Nielsen numbers. Proceedings of the International Conference on Topological methods in Nonlinear Analysis, Gdansk, 1997, 55-79.

[36] Fel'shtyn A.L. ,Hill R., Wong P, Reidemeister numbers of equivariant maps. Topology and its Applications vol. 67 (1995), 119 - 131.

[37] A.T.Fomenko, Symplectic geometry. Advan.Stud. in Con. Math., v.5 (1988), 1-388.

[38] J. Franks, Homology and dynamical systems. Regional Conf. Ser. Math, 49 (1982), 1-120.

[39] W. Franz, Über die Torsion einer Überdeckung. J. Reine Angew. Math., 173 (1935), 245-254.

[40] D. Fried, Periodic points and twisted coefficients. Lect. Notes in Math., 1007 (1983), 261-293.

[41] D. Fried, Lefschetz formula for flows, The Lefschetz centennial conference. Contemp. Math., 58 (1987), 19-69.

[42] D. Fried, Homological identities for closed orbits. Invent. Math., 71 (1983), 219-246.

[43] I.Gontareva,A.L.Fel'shtyn, An analogue of Morse inequalities for domain of attraction. Vestnik Leningrad.Univ.Math.v.17,n.2,1984,p.16-20.

[44] M. Gromov, Groups of polynomial growth and expanding maps. Publicationes Mathematiques, 53, 1981, 53-78.

[45] M. Gromov, Hyperbolic groups, Essays in Group theory. Mathematical Sciences Research Institute Publications, 8, 1987, 75-265.

[46] M.Handel,W.P.Thurston, New proofs of some results of Nielsen. Advan. in Math.,56(1985),2,p.267-289.

[47] M. Handel, The entropy of orientation reversing homeomorphisms of surfaces. Topology 21(1982), 291-296.

[48] P. R. Heath, Product formulae for Nielsen numbers of fibre maps. Pacific J. Math. 117 (2) (1985), 267-289.

[49] P. R. Heath, R. Piccinini, C. You, Nielsen-type numbers for periodic points I. Lecture Notes in Math. Vol. 1411 (1988) p.86-88.

[50] R. Hill, Some new results on Reidemeister numbers from group theoretical point of view. Preprint, 1993.

[51] B. Jiang, Lectures on Nielsen Fixed Point Theory. Contemp. Math. 14, AMS, 1983.

[52] B. Jiang, Estimation of the number of periodic orbits. Preprint of Universität Heidelberg, Mathematisches Institut, Heft 65, Mai 1993.

[53] B. Jiang, S. Wang, Lefschetz numbers and Nielsen numbers for homeomorphisms on aspherical manifolds. Topology - Hawaii, World Scientific , Singapore (1992), 119-136.

[54] J. Igusa, On the graded ring of theta-constants. Amer. J. Math., 86 (1964), 219-246.

[55] N. V. Ivanov, Entropy and the Nielsen Numbers. Dokl. Akad. Nauk SSSR 265 (2) (1982), 284-287 (in Russian); English transl.: Soviet Math. Dokl. 26 (1982), 63-66.

[56] N. V. Ivanov, Nielsen numbers of maps of surfaces. Journal Sov. Math., 26, (1984).

[57] A.A. Kirillov , Elements of the Theory of Representations, Springer Verlag 1976.

[58] T. Kobayashi, Links of homeomorphisms of surfaces and topological entropy. Proceed. of the Japan Acad. 60(1984), 381-383.

[59] K.Komia, Fixed point indices of equivariant maps and Möbius inversion. Invent. Math. 91(1988) , 129-135.

[60] S.Lang , Algebra, Addison-Wesley 1993.

[61] R. Lee, E. Miller, S. Weintraub, The Rochlin invariant,theta functions and the holonomy of some determinant line bundle. J. reine angew. Math., 392 (1988), 187-218.

[62] S. Lefschetz, Continuous transformations of manifolds. Proc. Nat. Acad. Sci. U.S.A. 9 (1923), 90-93.

[63] A. Mal'cev, On a class of homogeneous spaces. Izvestiya Akademii Nauk SSSR. Seriya Matematičeskaya, 13 (1949), 9-32.

[64] A. Manning, Axiom A diffeomorphisms have rational zeta function. Bull. London Math. Soc. 3 (1971), 215-220.

[65] J. Milnor, Infinite cyclic covers. Proc. Conf. "Topology of Manifolds" in Michigan 1967, 115-133.

[66] J. Milnor, A duality theorem for the Reidemeister torsion. Annals of Math., v.76(1962), 137-147.

[67] J.Milnor, W. Thurston, On iterated maps of the interval. Lec. Not. in Math., vol. 1342, 465-564.

[68] J.Milnor and A.Wallace, Differential Topology. 1972.

[69] W. Müller, Analytic torsion and R-torsion of Riemannian manifolds. Adv. in Math. 28 (1978), 233-305.

[70] J. Nielsen, Untersuchungen zur Topologie der geschlossenen zweiseiti-gen Flächen. Acta Math., 50 (1927), 189-348.

[71] B. Norton-Odenthal, Ph. D Thesis. University of Wisconsin, Madison, 1991.

[72] W. Parry and M. Pollicot, Zeta functions and the periodic structure of hyperbolic dynamics. Asterisque, vol.187-188, 1990

[73] S.J. Patterson, An introduction to the theory of the Riemann zeta-function. Cambridge studies in advanced mathematics 14 , 1988.

[74] V. B. Pilyugina and A. L. Fel'shtyn, The Nielsen zeta function. Funk-tsional. Anal. i Prilozhen. 19 (4) (1985), 61-67 (in Russian); English transl.: Functional Anal. Appl. 19 (1985), 300-305.

[75] F.Przytycki, An upper estimation for topological entropy. Inventiones Math., 59(1980), 205-213.

[76] D. Quillen, Determinants of Cauchy-Riemann operators over a Riemann surface. Functional Anal. i Prilozhen.,19(1)(1985),31-34.

[77] D. Ray and I. Singer, R-torsion and the Laplacian an Riemannian manifolds. Adv. in Math. 7 (1971), 145-210.

[78] K. Reidemeister, Automorphismen von Homotopiekettenringen. Math. Ann. 112 (1936), 586-593.

[79] G. de Rham, Complexes a automorphismes et homeomorphie differentiable. Ann. Inst. Fourier, 2 (1950), 51-67.

[80] V.A. Rohlin and D.B.Fuks, Introductory course in topology.Geometric chapters.Nauka, Moscow ,1977.

[81] W. Rudin , Fourier Analysis on Groups. Interscience tracts in pure and applied mathematics number 12, 1962.

[82] D. Ruelle, Zeta function for expanding maps and Anosov flows. Invent. Math, 34 (1976), 231- 242.

[83] D. Ruelle, An inequality for the entropy of differentiable maps. Bol. Soc. Brasil. Mat., 9(1978), 83-88.

[84] A. Schwarz, The partition function of degenerate quadratic functional and Ray-Singer torsion. Lett. Math. Phys., 2(1978), 247-252.

[85] Smooth dynamical systems(Russian translation). Mir,Moscow, 1977.

[86] M. Shub, Endomorphisms of compact differentiable manifolds. Amer. J. Math. 91 (1969), 175-179.

[87] S. Smale, Differentiable dynamical systems. Bull. Amer. Math. Soc. 73 (1967), 747-817.

[88] S.Smale, On gradient dynamical systems. Annals of Math., 74(1961), 199-206.

[89] H. Steinlein, Ein Satz über den Leray-Schauderschen Abbildungsgrad. Math.Zeit., 126(1972), 176-208.

[90] D. Sullivan, Travaux de Thurston sur les groupes quasi-Fuchsiens et les varietes hyperboliques de dimension 3 fibres sur S^1. Seminar Bourbaki 554(1979/80), 1-19.

[91] W. Thurston, On the geometry and dynamics of diffeomorphisms of surfaces. Bull. AMS 19 (1988), 417-431.

[92] W. Thurston, Existence of codimension one foliated manifolds. Annals of Math., v.104(1976), 249-268.

[93] W. Thurston, Hyperbolic structures on 3-manifolds, II: surface groups and 3-manifolds which fibers over the circle. Preprint.

[94] V. Turaev, The cohomology rings,formula for the linking coefficients and the spin structure invariant. Matem. Sbornik 120 (1983), 68-84.

[95] V. Turaev, Quantum Invariants of Knots and 3-manifolds. de Gruyter Studies in Mathematics 18, 1994.

[96] H.Ulrich, Fixed point theory of parametrized equivariant maps. Lecture Notes in Math. v.1343, 1988

[97] F.Wecken, Fixpunktklassen, II. Math. Ann. 118 (1942), 216-234.

[98] A. Weil, Numbers of solutions of equations in finite fields. Bull. AMS. 55 (1949), 497-508.

[99] F.W. Wilson, The structure of the level surfaces of a Laypunov function. Jour.Diff.Equat., 3(1967), 323-329.

[100] E. Witten, Global gravitational anomalies. Comm. Math. Phys., 100 (1985), 197-229.

[101] E. Witten, Quantum field theory and the Jones polynomial. Commun. Math. Phys., 121(1988), 351-399.

[102] P.P. Zabreiko, M.A. Krasnosel'skii, Iterations of operators and fixed
      points. Dokl.Akad. Nauk SSSR 196(1971), 1006-1009.

Institut für Mathematik,E.-M.-Arndt- Universität Greifswald
Jahn-strasse 15a, D-17489 Greifswald, Germany.
E-mail address: felshtyn@mail.uni-greifswald.de

Editorial Information

To be published in the *Memoirs*, a paper must be correct, new, nontrivial, and significant. Further, it must be well written and of interest to a substantial number of mathematicians. Piecemeal results, such as an inconclusive step toward an unproved major theorem or a minor variation on a known result, are in general not acceptable for publication. Papers appearing in *Memoirs* are generally longer than those appearing in *Transactions*, which shares the same editorial committee.

As of May 31, 2000, the backlog for this journal was approximately 7 volumes. This estimate is the result of dividing the number of manuscripts for this journal in the Providence office that have not yet gone to the printer on the above date by the average number of monographs per volume over the previous twelve months, reduced by the number of volumes published in four months (the time necessary for preparing a volume for the printer). (There are 6 volumes per year, each containing at least 4 numbers.)

A Consent to Publish and Copyright Agreement is required before a paper will be published in the *Memoirs*. After a paper is accepted for publication, the Providence office will send a Consent to Publish and Copyright Agreement to all authors of the paper. By submitting a paper to the *Memoirs*, authors certify that the results have not been submitted to nor are they under consideration for publication by another journal, conference proceedings, or similar publication.

Information for Authors

Memoirs are printed from camera copy fully prepared by the author. This means that the finished book will look exactly like the copy submitted.

The paper must contain a *descriptive title* and an *abstract* that summarizes the article in language suitable for workers in the general field (algebra, analysis, etc.). The *descriptive title* should be short, but informative; useless or vague phrases such as "some remarks about" or "concerning" should be avoided. The *abstract* should be at least one complete sentence, and at most 300 words. Included with the footnotes to the paper should be the 2000 *Mathematics Subject Classification* representing the primary and secondary subjects of the article. The classifications are accessible from www.ams.org/msc/. The list of classifications is also available in print starting with the 1999 annual index of *Mathematical Reviews*. The Mathematics Subject Classification footnote may be followed by a list of *key words and phrases* describing the subject matter of the article and taken from it. Journal abbreviations used in bibliographies are listed in the latest *Mathematical Reviews* annual index. The series abbreviations are also accessible from www.ams.org/publications/. To help in preparing and verifying references, the AMS offers MR Lookup, a Reference Tool for Linking, at www.ams.org/mrlookup/. When the manuscript is submitted, authors should supply the editor with electronic addresses if available. These will be printed after the postal address at the end of the article.

Electronically prepared manuscripts. The AMS encourages electronically prepared manuscripts, with a strong preference for $\mathcal{A}_{\mathcal{M}}\mathcal{S}$-LaTeX. To this end, the Society has prepared $\mathcal{A}_{\mathcal{M}}\mathcal{S}$-LaTeX author packages for each AMS publication. Author packages include instructions for preparing electronic manuscripts, the *AMS Author Handbook*, samples, and a style file that generates the particular design specifications of that publication series. Though $\mathcal{A}_{\mathcal{M}}\mathcal{S}$-LaTeX is the highly preferred format of TeX, author packages are also available in $\mathcal{A}_{\mathcal{M}}\mathcal{S}$-TeX.

Authors may retrieve an author package from e-MATH starting from `www.ams.org/tex/` or via FTP to `ftp.ams.org` (login as `anonymous`, enter username as password, and type `cd pub/author-info`). The *AMS Author Handbook* and the *Instruction Manual* are available in PDF format following the author packages link from `www.ams.org/tex/`. The author package can be obtained free of charge by sending email to `pub@ams.org` (Internet) or from the Publication Division, American Mathematical Society, P.O. Box 6248, Providence, RI 02940-6248. When requesting an author package, please specify $\mathcal{AMS}$-L^ATEX or $\mathcal{AMS}$-TEX, Macintosh or IBM (3.5) format, and the publication in which your paper will appear. Please be sure to include your complete mailing address.

Sending electronic files. After acceptance, the source file(s) should be sent to the Providence office (this includes any TEX source file, any graphics files, and the DVI or PostScript file).

Before sending the source file, be sure you have proofread your paper carefully. The files you send must be the EXACT files used to generate the proof copy that was accepted for publication. For all publications, authors are required to send a printed copy of their paper, which exactly matches the copy approved for publication, along with any graphics that will appear in the paper.

TEX files may be submitted by email, FTP, or on diskette. The DVI file(s) and PostScript files should be submitted only by FTP or on diskette unless they are encoded properly to submit through email. (DVI files are binary and PostScript files tend to be very large.)

Electronically prepared manuscripts can be sent via email to `pub-submit@ams.org` (Internet). The subject line of the message should include the publication code to identify it as a Memoir. TEX source files, DVI files, and PostScript files can be transferred over the Internet by FTP to the Internet node `e-math.ams.org` (130.44.1.100).

Electronic graphics. Comprehensive instructions on preparing graphics are available at `www.ams.org/jourhtml/graphics.html`. A few of the major requirements are given here.

Submit files for graphics as EPS (Encapsulated PostScript) files. This includes graphics originated via a graphics application as well as scanned photographs or other computer-generated images. If this is not possible, TIFF files are acceptable as long as they can be opened in Adobe Photoshop or Illustrator. No matter what method was used to produce the graphic, it is necessary to provide a paper copy to the AMS.

Authors using graphics packages for the creation of electronic art should also avoid the use of any lines thinner than 0.5 points in width. Many graphics packages allow the user to specify a "hairline" for a very thin line. Hairlines often look acceptable when proofed on a typical laser printer. However, when produced on a high-resolution laser imagesetter, hairlines become nearly invisible and will be lost entirely in the final printing process.

Screens should be set to values between 15% and 85%. Screens which fall outside of this range are too light or too dark to print correctly. Variations of screens within a graphic should be no less than 10%.

Inquiries. Any inquiries concerning a paper that has been accepted for publication should be sent directly to the Electronic Prepress Department, American Mathematical Society, P. O. Box 6248, Providence, RI 02940-6248.

Selected Titles in This Series

(*Continued from the front of this publication*)

For a complete list of titles in this series, visit the
AMS Bookstore at **www.ams.org/bookstore/**.